QUANTITATIVE METHODS AND ANALYTICAL TECHNIQUES IN FOOD MICROBIOLOGY

Challenges and Health Implications

QUANTITATIVE METHODS AND ANALYTICAL TECHNIQUES IN FOOD MICROBIOLOGY

Challenges and Health Implications

Edited by
Leonardo Sepúlveda Torre, PhD
Cristóbal Noé Aguilar, PhD
Porteen Kannan, PhD
A. K. Haghi, PhD

AAP APPLE
ACADEMIC
PRESS

First edition published 2022

Apple Academic Press Inc.
1265 Goldenrod Circle, NE,
Palm Bay, FL 32905 USA
4164 Lakeshore Road, Burlington,
ON, L7L 1A4 Canada

CRC Press
6000 Broken Sound Parkway NW,
Suite 300, Boca Raton, FL 33487-2742 USA
2 Park Square, Milton Park,
Abingdon, Oxon, OX14 4RN UK

© 2022 Apple Academic Press, Inc.

Apple Academic Press exclusively co-publishes with CRC Press, an imprint of Taylor & Francis Group, LLC

Library and Archives Canada Cataloguing in Publication

Title: Quantitative methods and analytical techniques in food microbiology : challenges and health implications / edited by Leonardo Sepúlveda Torre, PhD, Cristóbal Noé Aguilar, PhD, Porteen Kannan, PhD, A. K. Haghi, PhD.
Names: Sepúlveda Torre, Leonardo, editor. | Aguilar, Cristóbal Noé, editor. | Kannan, Porteen, editor. | Haghi, A. K., editor.
Description: First edition. | Includes bibliographical references and index.
Identifiers: Canadiana (print) 2021039336X | Canadiana (ebook) 20210393394 | ISBN 9781774637265 (hardcover) | ISBN 9781774637425 (softcover) | ISBN 9781003277453 (ebook)
Subjects: LCSH: Food—Microbiology. | LCSH: Foodborne diseases—Microbiology. | LCSH: Food—Safety measures.
Classification: LCC QR115 .Q83 2022 | DDC 664.001/579—dc23

Library of Congress Cataloging-in-Publication Data

....

CIP data on file with US Library of Congress

....

ISBN: 978-1-77463-726-5 (hbk)
ISBN: 978-1-77463-742-5 (pbk)
ISBN: 978-1-00327-745-3 (ebk)

About the Editors

Leonardo Sepúlveda Torre, PhD
*Full Professor, Bioprocesses and Microbial Biochemistry Group,
School of Chemistry, Autonomous University of Coahuila, Saltillo,
Coahuila, México*

Leonardo Sepúlveda Torre, PhD, is a chemist specializing in Bromatology at the School of Chemistry of the Autonomous University of Coahuila (UAdeC). His postgraduate studies were in related topics on Food Biotechnology in the Food Research Department of the UAdeC. In 2011 he made a research stay at the "Institute of Biotechnology and Bioengineering at the University of Minho (Uminho), Braga, Portugal. He worked as a collaborator at the Center of Biological Engineering" in Uminho, Braga, Portugal, with the link project "Biotechnologies for regional food biodiversity in Latin America." At the same institution, he carried out his postdoctoral stay in 2015–2016 on "Assisted extraction by fermentation of polyphenols from agro-industrial waste." He is currently a professor-researcher in the School of Chemistry of the UAdeC, responsible for the group Bioprocesses and Microbial Biochemistry. He teaches classes in algebra, differential calculus, differential equations, accounting, and financial administration, general microbiology, food microbiology in chemical engineering, chemical, and QFB programs.

Cristóbal Noé Aguilar, PhD
*Full Professor, Associate Editor of Heliyon (Microbiology) and Frontiers
in Sustainable Food Systems (Food Processing), Bioprocesses and
Bioproducts Research Group, Food Research Department, School of
Chemistry, Autonomous University of Coahuila, Saltillo, Mexico*

Cristóbal Noé Aguilar, PhD, is a Director of Research and Postgraduate Programs at the Universidad Autonoma de Coahuila, Mexico. Dr. Aguilar has published more than 160 papers in indexed journals, more than 40 articles in Mexican journals, and 250 contributions in scientific meetings. He has also published many book chapters, several Mexican books, four editions of international books, and more. He has been awarded several prizes and awards, the most important of which are the National Prize of Research (2010) from the Mexican Academy of Sciences; the Prize "Carlos Casas Campillo (2008)" from the Mexican Society of Biotechnology and Bioengineering; National Prize AgroBio-2005; and the Mexican Prize in Food Science and Technology. Dr. Aguilar is a member of the Mexican Academy of Science, the International Bioprocessing Association, the Mexican Academy of Sciences, the Mexican Society for Biotechnology and Bioengineering, and the Mexican Association for Food Science and Biotechnology. He has developed more than 21 research projects, including six international exchange projects.

Porteen Kannan, PhD
Assistant Professor, Department of Veterinary Public Health,
Madras Veterinary College, Tamil Nadu Veterinary and Animal Sciences
University, India

Porteen Kannan, PhD, is an Assistant Professor in the Department of Veterinary Public Health at Madras Veterinary College, Tamil Nadu Veterinary and Animal Sciences University, India. The research activities of Dr. Kannan include food safety and anti-microbial resistance. He performed his postdoctoral studies at the US Department of Agriculture, Maryland, USA, with a specialization in foodborne pathogens. He has published his work in both national and international journals. He is actively involved in mentoring both MVSc and PhD students.

A. K. Haghi, PhD
Professor Emeritus of Engineering Sciences, Former Editor-in-Chief,
International Journal of Chemoinformatics and Chemical Engineering
and Polymers Research Journal; Member, Canadian Research and
Development Center of Sciences and Culture

A. K. Haghi, PhD, is the author and editor of 200 books, as well as over 1,000 published papers in various journals and conference proceedings. Dr. Haghi has received several grants, consulted for a number of major corporations, and is a frequent speaker to national and international audiences. Since 1983, he has served as a professor at several universities. He is former Editor-in-Chief of the *International Journal of Chemoinformatics and Chemical Engineering* and *Polymers Research Journal*, and is on the editorial boards of many international journals. He is also a member of the Canadian Research and Development Center of Sciences and Cultures (CRDCSC), Montreal, Quebec, Canada. He holds a BSc in urban and environmental engineering from the University of North Carolina (USA), an MSc in mechanical engineering from North Carolina A&T State University (USA), a DEA in applied mechanics, acoustics, and materials from the Université de Technologie de Compiègne (France), and a PhD in engineering sciences from Université de Franche-Comté (France).

Contents

Contributors

Cristóbal Noé Aguilar
Bioprocesses and Bioproducts Research Group, Food Research Department, School of Chemistry,
Autonomous University of Coahuila, Saltillo Unit – 25280, Coahuila, México,
E-mail: cristobal.aguilar@uadec.edu.mx

Shayma Thyab Gddoa Al-Sahlany
Department of Food Science, College of Agriculture, University of Basrah, Basra City, Iraq

J. A. Ascacio-Valdés
Bioprocesses and Bioproducts Research Group, Food Research Department, School of Chemistry,
Autonomous University of Coahuila, Saltillo – 25280, Coahuila, México,
E-mail: alberto_ascaciovaldes@uadec.edu.mx

Kolawole Banwo
Food Microbiology and Biotechnology Unit, Department of Microbiology, University of Ibadan,
Oyo State, Nigeria

Jose J. Buenrostro-Figueroa
Research Center for Food and Development A.C., Cd. Delicias – 33088, Chihuahua, México

Débora A. Campos
Universidade Católica Portuguesa, CBQF-Centro de Biotecnologia e Química Fina-Laboratório
Associado, Escola Superior de Biotecnologia, Rua Diogo Botelho 1327, Porto – 4169-005, Portugal

Nadia D. Cerda-Cejudo
Bioprocesses and Bioproducts Research Group, Food Research Department, School of Chemistry,
Autonomous University of Coahuila, Saltillo – 25280, Coahuila, México

Mónica L. Chávez-González
Bioprocesses and Bioproducts Research Group, Food Research Department, School of Chemistry,
Autonomous University of Coahuila, Saltillo – 25280, Coahuila, México,
E-mail: monicachavez@uadec.edu.mx

Adriana C. Flores-Gallegos
Bioprocesses and Bioproducts Research Group, Food Research Department,
Autonomous University of Coahuila, Saltillo, México, E-mail: carolinaflores@uadec.edu.mx

Andrea Guadalupe Flores-Valdés
Food Research Department, School of Chemistry, Autonomous University of Coahuila,
Saltillo – 25280, Coahuila, México

Alfredo Ivanoe García-Galindo
Center for Interdisciplinary Studies and Research, Autonomous University of Coahuila,
Saltillo, México

Ricardo Gómez-García
Universidade Católica Portuguesa, CBQF-Centro de Biotecnologia e Química Fina-Laboratório
Associado, Escola Superior de Biotecnologia, Rua Diogo Botelho 1327, Porto – 4169-005, Portugal

Nathiely Ramírez Guzmán
Center for Interdisciplinary Studies and Research, Autonomous University of Coahuila,
Saltillo – 25020, Coahuila, México

Anna Iliná
Food Research Department, School of Chemistry, Autonomous University of Coahuila,
Saltillo – 25280, Coahuila, México

Anna Ilyina
Bioprocesses and Bioproducts Research Group, Food Research Department, School of Chemistry,
Autonomous University of Coahuila, C.P. – 25280, Saltillo, Coahuila, México

Ana R. Madureira
Universidade Católica Portuguesa, CBQF-Centro de Biotecnologia e Química Fina-Laboratório
Associado, Escola Superior de Biotecnologia, Rua Diogo Botelho 1327, Porto – 4169-005, Portugal

José Luis Martínez-Hernández
Bioprocesses and Bioproducts Group, Food Research Department, Autonomous University of
Coahuila, Saltillo, México; Nanobioscience Group, School of Chemistry, Autonomous University of
Coahuila, C.P. – 25280, Saltillo, Coahuila, México, E-mail: jose-martinez@uadec.edu.mx

Gloria A. Martínez-Medina
Bioprocesses and Bioproducts Research Group, Food Research Department, School of Chemistry,
Autonomous University of Coahuila, Saltillo – 25280, Coahuila, México

Georgina Michelena-Álvarez
Cuban Institute for Research on Sugarcane Derivates, Postal Zone 10, San Miguel del Padrón,
La Habana City, Cuba

Balaram Mohapatra
Department of Bioscience and Bioengineering, Indian Institute of Technology Bombay,
Bombay – 400076, Maharashtra, India

Lorena Moreno-Vilet
Department of Food Technology, CONACYT-Department of Food Technology, Centre of Research
and Assistance in Technology and Design of the State of Jalisco, A.C. Zapopan, Jalisco, México

Alberto A. Neira-Vielma
Nanobioscience Group, School of Chemistry, Autonomous University of Coahuila, C.P. – 25280,
Saltillo, Coahuila, México, E-mail: aneiravielma@uadec.edu.mx

Alaa Kareem Niamah
Department of Food Science, College of Agriculture, University of Basrah, Basra City, Iraq,
E-mails: alaakareem2002@hotmail.com; alaa.niamah@uobasrah.edu.iq

Ami R. Patel
Division of Dairy and Food Microbiology, Mansinhbhai Institute of Dairy and Food Technology-
MIDFT, Dudhsagar Dairy Campus, Mehsana – 384-002, Gujarat, India

Manuela Pintado
Universidade Católica Portuguesa, CBQF-Centro de Biotecnologia e Química Fina-Laboratório
Associado, Escola Superior de Biotecnologia, Rua Diogo Botelho 1327, Porto – 4169-005, Portugal,
E-mail: mpintado@porto.ucp.pt

Nathiely Ramirez-Guzman
Center for Interdisciplinary Studies and Research, Autonomous University of Coahuila, Saltillo,
México, E-mail: nathiely.ramirez@uadec.edu.mx

Josefina Rodríguez
Center for Interdisciplinary Studies and Research, Autonomous University of Coahuila, Saltillo, México

Raúl Rodríguez-Herrera
Bioprocesses and Bioproducts Research Group, Food Research Department, Universidad Autonoma de Coahuila, Saltillo, Mexico, E-mail: raul.rodriguez@uadec.edu.mx

José Sandoval-Cortes
Analytical Chemistry Department, School of Chemistry, Autonomous University of Coahuila , Saltillo, México, E-mail: josesandoval@uadec.edu.mx

Leonardo Sepúlveda
Bioprocesses and Bioproducts Research Group, Food Research Department, School of Chemistry, Autonomous University of Coahuila, Saltillo – 25280, Coahuila, México

Smita Singh
Department of Life Sciences (Food Technology), Graphic Era (Deemed to be) University, Dehradun, Uttarakhand – 248002, India

Mamta Thakur
Department of Food Engineering and Technology, Sant Longowal Institute of Engineering and Technology, Longowal – 148106, Punjab, India

Cristian Torres-León
Research Center and Ethnobiological Garden, Autonomous University of Coahuila, Viesca – 27480, Coahuila, México, E-mail: ctorresleon@uadec.edu.mx

C. Guillermo Valdivia-Nájar
Centre of Research and Assistance in Technology and Design of the State of Jalisco, A.C. Zapopan, Jalisco, México, E-mail: gvaldivia@ciatej.mx

Deepak Kumar Verma
Department of Agricultural and Food Engineering, Indian Institute of Technology Kharagpur, Kharagpur – 721-302, West Bengal, India; Bioprocesses and Bioproducts Research Group, Food Research Department, School of Chemistry, Autonomous University of Coahuila, Saltillo Unit – 25280, Coahuila, México, E-mail: rajadkv@rediffmail.com

Abbreviations

AP	active packaging
ArGa	arabinogalactan
ATCC	American type culture collection
ATP	adenosine triphosphate
AXOS	arabinoxylo-oligosaccharides
BIA	biospecific interaction analysis
BOD	biochemical oxygen demand
CBHI	cellobiohydrolases I
CE	capillary electrophoresis
CP	cold plasma
DAF	DNA amplification fingerprinting
DF	dietary fiber
DGGE	denaturing gradient gel electrophoresis
FAO	Food and Agriculture Organization
FDA	Food and Drug Administration
FIR	far IR radiation
FISH	fluorescence in situ hybridization
FOS	fructooligosaccharides
FRET	fluorescent resonance energy transfer
GM	genetically modified
GMOs	genetically modified organisms
GOS	galactooligosaccharides
GRAS	generally recognized as safe
HAV	hepatitis A virus
HHDP	hexahydroxydiphenoyl
HMF	5-hydroxy methyl furfural
HPP	high pressure processing
HTST	high temperature short time
IBD	inflammatory bowel disease
IDF	insoluble dietary fiber
IOR	ionizing radiation
IR	infrared
ISM	industrial, scientific, and medical

ITS	internal transcribed spacer
LAB	lactic acid bacteria
L-Araf	L-Arabinofuranosyl
LED	light-emitting diodes
LiP	lignin-peroxidase
MIR	medium IR radiation
MnP	manganese peroxidase
MW	microwaves
NIR	near IR radiation
NoV	norovirus
NTT	nonthermal technologies
OH	ohmic heating
PCR	polymerase chain reaction
PEF	pulsed electric fields
PFBs	plant food by-products
PL and UV	pulsed light and UV processing
PL	pulsed light
PNAs	peptide nucleic acids
PP	passive packaging
PPO	polyphenol oxidase
qPCR	quantitative PCR
RF	radiofrequency
RFLP	restriction fragment length polymorphism
RT-PCR	real-time PCR
SCAR	sequence characterized by amplified region
SCFA	short chain fatty acids
SCP	single cell protein
SDF	soluble dietary fiber
SMF	submerged fermentation
SSF	solid-state fermentation
TGGE	temperature gradient gel electrophoresis
T-RFLP	terminal restriction fragment length polymorphism
US	ultrasound
UTH	ultra-high temperature
UV	ultraviolet
Vis	visible
WA	water activity

Preface

Since ancient times, human society has been closely related to the ability to acquire enough food, so that not only the basic needs of survival were met, but also the preservation of food place allowed humans to be able to devote time to the arts, crafts, and sciences.

The development of one of the oldest activities of human beings, agriculture, is closely linked to the ability to preserve food, first by techniques developed over the centuries by trial-and-error observation, and more recently by the increasing application of science and engineering. The basis of these advances is our knowledge of food microbiology. Long before, Antoine van Leeuwenhoek described his living animalcules, many of the conditions that controlled microbial deterioration had been empirically identified. However, it was the emergence of microbiology science that promoted the preservation of food from an art to a science, allowing food to be processed, distributed, and marketed with a high degree of confidence in terms of product quality and safety. Thus, food microbiology has been an important part of the discipline since its early days.

The scope of food microbiology is highly inclusive, as it interacts with all subdisciplines of microbiology (e.g., public health microbiology, microbial genetics, fermentation technologies, microbial physiology, and biochemistry). In addition, food microbiologists have been at the forefront of many microbiological concepts and advances. For example, the development of biofilms and the ability to detect low numbers of metabolically stressed microbes from highly complex matrices are two areas where food microbiologists are providing critical insights into the behavior of microbiological systems. In addition, new research topics have been raised as a result of the unique challenge given to food microbiologists, such as predictive microbiology, probiotics, microbial risk assessments, and natural antimicrobials.

This book has been prepared to provide up-to-date and detailed scientific information on food microbiology. The book is organized into 15 chapters, five of which focus on the two fundamental aspects of the matter, food, and microorganisms. Each section consists of detailed information, from the generalities to the particular aspects of each topic of relevance to be addressed, including basic microbiology, safety, food, pathogenic microorganisms, food conservation, sanitization, hygiene procedures, etc.

The microbial diversity found in food is described from the classification by kingdoms and the main groups of microorganisms present in them.

Although the main issue is about microbial food pathogens, the book also covers another important aspect of food microbiology, such as food systems and measures to prevent and control food, foodborne diseases, etc. Uncontrolled and unwanted microbial growth destroys large quantities of food age, causing significant losses both economically and relative to nutrient content. In addition, the consumption of foods contaminated with particular microorganisms or microbial products can also cause serious illnesses, such as food-mediated infections and food poisoning. Every minute, there are about 50,000 cases of gastrointestinal diseases, and many individuals, especially children, die from these infections. The most important preventive measures are for the development and continuous implementation of effective interventions to improve overall food safety.

—Editors

CHAPTER 1

Classification of Microorganisms and Food Microbiology Generalities

ALFREDO IVANOE GARCÍA-GALINDO,[1] LEONARDO SEPÚLVEDA,[2] and
CRISTÓBAL NOÉ AGUILAR[2]

[1]*Center for Interdisciplinary Studies and Research,
Autonomous University of Coahuila, Saltillo, México*

[2]*Bioprocesses and Bioproducts Research Group, Food Research
Department, Autonomous University of Coahuila, Saltillo, México,
E-mail: cristobal.aguilar@uadec.edu.mx (C. N. Aguilar)*

ABSTRACT

Each life form is part of one of the five known kingdoms: Plantae, Animalia, Protista, Fungi, and Monera. The structural unit of life is the cell, in which genetic information is replicated and stored, cellular components are synthesized, and energy is generated, also there are cells which have mechanisms for movement. Plants and animals are multicellular beings. Plants have cell walls composed of cellulose, hemicellulose, pectin, and lignin, while animal cells have no cell wall. In plants, the energy process known as photosynthesis is carried out, while animals must ingest and digest food. To perform each specific function, the cellular material of each plant and animal is differentiated into tissues, organs, and systems. All members of the protist kingdom are unicellular forms, which perform the functions with no specialization, however, some members of this kingdom form well-developed aggregates (colonies) that reassemble multicellular structures. The fungal kingdom is formed by unicellular (yeast) and multicellular structures or filamentous masses (filamentous fungi). Unicellular life forms belong to the Monera kingdom, such as bacteria and blue green algae [1].

These five kingdoms can be divided into two groups: eukaryotes (protists, fungi, animals, and plants) and prokaryotes (Monera). The two groups differ by the nucleus. In eukaryotic cells, the nucleus is a discrete body surrounded by a membrane. The nucleus is divided according to the classic Mendelian description. Each cell has vacuoles, Golgi apparatus, and mitochondria. Some plant cells and protists also have plastids [2]. Prokaryotic cells do not contain defined structures of the organelles, their nucleus is diffuse, sectioned, and has no membrane. During the cell division, the nucleus separates into two parts, each capable of generating a new cell. Within prokaryotes, the organisms of greatest interest to the food microbiologist are bacteria, while eukaryotes are fungi [3].

1.1   FUNGI

Multicellular fungi are known as filamentous fungi, while unicellular fungi as yeasts. However, there are dimorphic microorganisms capable of growing as unicellular or multicellular, and their growth strongly depends on the conditions and culture time. Many fungi grow by accommodating their cells one after the other, forming filaments or hyphae, which branch out and together form the mycelium. Some fungi can gather their hyphae so much that they form firm structures called fruiting bodies, like mushrooms. Some fungi have divisions between each cell, while others do not, the former are known as septate, while the latter are known as non-septate [4].

Fungi differed by their ways of reproduction. All fungi, except those belonging to the Mycelia Sterila class, can produce spores asexually by simple cell division without nucleus fusion. Members of the Phycomycetes, Basidiomycetes, and Ascomycetes classes also reproduce sexually by fusion of two similar or non-similar cells, but asexual spores occur more than sexual. Because it is extremely difficult to induce forms of sexual or perfect reproduction, they are considered within a class called Fungi imperfecti or deuteromycetes [5].

1.1.1   PHYCOMYCETES

This class of fungi is divided into three subclasses: Archimycetes, Oomycetes, and Zygomycetes. Members of the Archimycetes subclass are primitive aquatic fungi, which are not of considerable relevance to food microbiologists. The members of the Oomycetes subclass are also considered primitive aquatic fungi; however, these can be isolated from soil and some fruits where they can rot. The most important members of this subclass are the genera: Pythium, Saprolegnia, Phytophthora, and Plasmopara. A large part of the fungi of interest in food microbiology belong to the Zygomycetes subclass and the genera Mucor, Rhizopus, Mortierella, Thamnidium, Choanephora, Cunninghamella, and Syncephalastrum, capable of growing in fruits, vegetables, meats, bread, and cereals [6].

1.1.2 BASIDIOMYCETES

Some members of the basidiomycetes class are of great importance in the food area, since they form fruiting bodies that are generally edible, such is the case of Ustilaginales, Tremalla, and Urocystales. Examples of them are fungi commonly known as mushrooms and huitlacoches [7].

1.1.3 ASCOMYCETES

The class with the highest number of members is the Ascomycetes, comprising yeasts to filamentous fungi capable of forming fruiting bodies. Ascomycetes do not have much importance for the microbial in food [8].

1.1.4 FUNGI IMPERFECTI

The members of the fungi imperfecti class are a special group of fungi for food microbiologists, many of them produce primary or secondary metabolites of industrial interest, and because many of them are pathogenic to humans through food consumption. The genera Aspergillus, Penicillium, Fusarium, Talaromyces, and Gibberella belong to this class [9].

1.1.5 MYCELIA STERILIA

The members of this class live in soils, generally of rice. They are not important for food microbiologists, but it is important to note that some members of the class Fungi imperfecti such as Fusarium and Helmintho-sporium, when losing their ability to produce conidiospores, erroneously they are placed within the Mycelia Sterila class [10].

1.2   CHARACTERISTICS OF FILAMENTOUS FUNGI

Filamentous fungi grow best on solid surfaces, the granular form of soil causes fungi to grow with a very branched filament. On food, filamentous fungi grow invasively without allowing other microorganisms to grow. The filamentous fungi associated with food are: Aspergillus, Penicillium, Rhizopus, Mucor, Absidia, Aurebasidium, Epiccocum, Alternaria, Helminthosporium, Monilia, Botrytis, Scopulariopsis, Cephalosporium, and Fusarium. Each of these fungi grows on defined types of food and under specific conditions [11].

1.3   YEAST

True yeasts are single-celled organisms belonging to the Ascomycetes and Fungi imperfecti classes. They have spherical, ellipsoidal, or elongated forms. Some produce a pseudomycelium and usually reproduce by budding. Yeasts are the simplest eukaryotic microorganisms, and much of their metabolism has been studied. Yeasts are widely used in the food industry, mainly in the production of bread and beer, and in the production of organic acids, enzymes, ethyl alcohol, ketone, and biomass. The most important genera are Saccharomyces, Torulopsis, and Candida [12].

1.4   BACTERIA

The Monera kingdom comprise Bacteria and Cyanobacteria, which are prokaryotes. Their cells have a three-dimensionally linked peptidoglycan cell wall, which produced very compact rigid structures. The amount of peptidoglycan and lipopolysaccharide material is used to taxonomically classify bacteria. Bacteria have a membrane, and, in some cases, they have a capsule structure. All bacteria associated with food are heterotrophic, they hydrolyze carbohydrates, proteins, lipids, and other food ingredients [13].

1.5 MICROORGANISMS AS FOOD PRODUCERS

Since the most remote historical times, microorganisms have been used to produce food. Microbial processes give rise to alterations in them that give them more resistance to deterioration or more desirable organoleptic characteristics like taste and texture. A lot of food manufacturing processes involving microorganisms are based on fermentation processes, mainly lactic fermentation of starting materials. This fermentation is usually carried out by lactic bacteria [14]. As a result, the pH decreases reducing the survival of undesirable bacterial species (mainly enteric bacteria), organic acids of short chain accumulate in the food that, in addition to their antibacterial effect, confer a pleasant taste, and, in certain cases, antibacterial compounds accumulate reducing the microbial food load by increasing its half-life or prevent the germination of spores of possible Gram-positive bacteria that cause food poisoning.

1.6 MICROORGANISMS AS FOOD SPOILAGE AGENTS

Impaired food is food damaged by microbial, chemical, or physical agents that made them unacceptable for human consumption. Food deterioration can cause important economic losses, approximately 20% of the fruits and vegetables collected are lost because of microbial deterioration caused by any of the 250 market diseases. The causative agents of deterioration can be bacteria, molds, and yeasts; being bacteria and molds the most important. Of all the microorganisms present in food, only some are able to multiply on the food, so being selected over time so that the initial heterogeneous population present in the food, is being reduced to more homogeneous populations and finally to an only type of microorganisms that manage to colonize all the food displacing others [15]. Therefore, during the deterioration process, a predominant population or type of microorganisms is selected so that the initial variety indicates little deterioration and reflects the initial populations.

There are several factors that determine the resistance to colonize food and these factors are:

1. **Intrinsic Factors:** They are the derivatives of the food composition: water activity (aw), pH, redox potential, nutrients, food structure, and antimicrobial agents present.

2. **Technological Treatments:** Factors that modify the initial flora because of food processing.
3. **Extrinsic Factors:** These derived from the physical conditions of the environment in which the food is stored.
4. **Implicit Factors:** The relationships established between microorganisms because of factors above [16].

Food can be attacked by a variety of microorganisms. Thus, each food type is damaged by the action of a specific type of microorganism, establishing a specific association between the altering microorganism and the altered product, for example, meat is the most easily deteriorated food due to favorable conditions for the growth of microorganisms derived from the aforementioned factors.

1.7 MICROORGANISMS AS FOODBORNE PATHOGENS

Certain pathogenic microorganisms are potentially transmissible through food. In these cases, the pathologies are usually gastrointestinal, although they can lead to more widespread symptoms in the body and even septicemia. The food pathologies can appear as isolated cases when a bad food processing has occurred, but they are usually associated with epidemic outbreaks more or less widespread in the territory; for example, the number of epidemic outbreaks associated with food in the last years throughout the national territory has oscillated between 900 and 1,000 outbreaks per year. The pathologies associated with food transmission are food infections caused by the ingestion of microorganisms or food poisoning produced as a result of ingestion of bacterial toxins produced by microorganisms present in the food. In certain cases, food allergies caused by the presence of microorganisms can occur [17].

In any case, for a toxin-infection to occur, the microorganism must produce:

- Enough number to colonize the intestine;
- Enough number to poison the intestine;
- Significant amounts of toxins.

Pathogenic microorganisms that produce food infections include bacteria, protozoa, and viruses, and food poisoning is caused by bacteria and fungi (molds). Bacteria to cause an infection, besides the above conditions it is necessary that the microorganism has a growth temperature range compatible with the body temperature of higher organisms (40°C). This is the reason that plant pathogens are not animal pathogens and that most psychrophiles and psychotrophs are not of great relevance in pathology [18]. A virus will be pathogenic only in the case that animal cells present the necessary receptors so that the virus can adsorb to them. Therefore, there is kingdom specificity between animal, plant, and bacterial viruses without cross-infection between kingdoms [19]. The origin of the pathogenic microorganism is classified in endogenous microorganisms present inside the food, and exogenous microorganisms deposited on the food surface. The former usually associated with animal foods since animal pathogens can be human, while plant pathogens cannot because differences between both types of microorganisms.

Finally, due to the importance of food poisoning in public health, the work of a food microbiologist is directed, in many cases, to the control and avoid the consumption of products made in poor conditions and, therefore, potentially dangerous. For this, a microbiological analysis must consider [20]:

- Sources of food contamination;
- Pathways of pathogen infection;
- The pathogens resistance to adverse conditions;
- The pathogen's growth needs;
- Minimize contamination and growth of microorganisms;
- Detection and isolation techniques;
- Sampling method proportional to the risk.

All the above requires the legal regulation of the microbiological characteristics of each food, which includes the definition of each food or food product and regulations on the tolerance of the number of permissible microorganisms (the so-called reference values).

KEYWORDS

- eukaryotes
- fungi
- microbiologist
- Monera
- multicellular structures
- nucleus

REFERENCES

1. Singh, B., & Satyanarayana, T., (2017). Basic microbiology. In: *Current Developments in Biotechnology and Bioengineering* (pp. 1–31). Elsevier: Netherlands.
2. Kim, G. C., (2013). Microbiology. In: *Bioprocess Engineering* (pp. 7–24). Woodhead Publishing: United Kingdom.
3. Killham, K., & Prosser, J. I., (2007). The prokaryotes. In: *Soil Microbiology, Ecology and Biochemistry* (pp. 119–144). Academic Press: USA.
4. Thorn, R. G., & Lynch, M. D. J., (2007). Fungi and eukaryotic algae. In: *Soil Microbiology, Ecology and Biochemistry* (pp. 145–162). Academic Press: USA.
5. Berman, J. J., (2019). Fungi. In: *Taxonomic Guide to Infectious Diseases* (pp. 229–262). Academic Press: United Kingdom.
6. Benekele, E. S., (2017). Fungi. In: *Handbook of Microbiology. Condensed Edition* (p. 940). CRC Press: USA.
7. Taylor, T. N., Krings, M., & Taylor, E. L., (2015). *Basidiomycota*. In: *Fossil Fungi* (pp. 173–199). Academic Press: United Kingdom.
8. Berman, J. J., (2012). *Ascomycota*. In: *Taxonomic Guide to Infectious Diseases* (pp. 199–208). Academic Press: United Kingdom.
9. Zabel, R. A., & Morrell, J. J., (2020). The characteristics and classification of fungi and bacteria. In: *Wood Microbiology* (pp. 55–98). Academic Press: United Kingdom.
10. Rana, K. L., Kour, D., Kaur, T., Devi, R., Negi, C., Yadav, A. N., Yadav, N., Singh, K., & Saxena, A. K., (2020). Endophytic fungi from medicinal plants: Biodiversity and biotechnological applications. In: *Microbial Endophytes* (pp. 273–305). Woodhead Publishing: United Kingdom.
11. El-Enshasy, H. A., (2007). Filamentous fungal cultures - process characteristics, products, and applications. In: *Bioprocessing for Value-Added Products from Renewable Resources* (pp. 225–261). Elsevier: USA.
12. Perricone, M., Gallo, M., Corbo, M. R., Sinigaglia, M., & Bevilacqua, A., (2017). Yeasts. In: *The Microbiological Quality of Food* (pp. 121–131). Woodhead Publishing: United Kingdom.
13. Philips, J. A., & Blaser, M. J., (2015). Introduction to bacteria and bacterial diseases. In: *Mandell, Douglas, and Bennett's Principles and Practice of Infectious Diseases* (pp. 2234–2236). Elsevier: USA.
14. Jackson, R. S., (2020). Fermentation. In: *Food Science and Technology, Wine Science* (pp. 461–572). Academic Press: United Kingdom.
15. Dilmaçünal, T., & Kuleaşan, H., (2018). Novel strategies for the reduction of microbial degradation of foods. In: *Food Safety and Preservation* (pp. 481–520). *Academic Press*: United Kingdom.
16. Majumdar, A., Pradhan, N., Sadasivan, J., Acharya, A., Ojha, N., Babu, S., & Bose, S., (2018). Food degradation and foodborne diseases: A microbial approach. In: *Handbook of Food Bioengineering, Microbial Contamination and Food Degradation* (pp. 109–148). Academic Press: United Kingdom.
17. Taylor, S. L., (2017). Disease processes in foodborne illness. In: *Foodborne Diseases* (pp. 3–30). Academic Press: United Kingdom.

18. Sharif, M. K., Javed, K., & Nasir, A., (2018). Foodborne illness: Threats and control. In: *Handbook of Food Bioengineering, Foodborne Diseases* (pp. 501–523). Academic Press: United Kingdom.

19. Maunula, L., & Von, B. C. H., (2016). Foodborne viruses in ready-to-eat foods. *Food Hygiene and Toxicology in Ready-to-Eat Foods* (pp. 51–68). Academic Press: United Kingdom.

20. Mendonca, A., Thomas-Popo, E., & Gordon, A., (2020). Microbiological considerations in food safety and quality systems implementation. In: *Food Safety and Quality Systems in Developing Countries* (pp. 185–260). Academic Press: United Kingdom.

CHAPTER 2

Risk and Safety in Microbiology

JOSEFINA RODRÍGUEZ[1] and CRISTÓBAL NOÉ AGUILAR[2]

[1]*Center for Interdisciplinary Studies and Research, Autonomous University of Coahuila, Saltillo, México*

[2]*Bioprocesses and Bioproducts Research Group, Food Research Department, Autonomous University of Coahuila, Saltillo, México, E-mail: cristobal.aguilar@uadec.edu.mx*

ABSTRACT

Over 4,000 infections associated with the laboratory practices have been reported in this century [1, 2]. Probably more have not been reported. Most of the victims work in research laboratories with microorganisms whose pathogenic potential was unknown at that time and/or used techniques that are currently dangerous. As a result of last 25 years of scientific research, the causes of these infections have been discovered and methods have been developed to prevent them, today, it is not necessary for those who work in microbiology laboratories (with certain exceptions that fall outside the objective of this book) to feel in danger of becoming infected with microorganisms when they work under safety standards and consider:

- The potential risks of the microorganisms in question;
- The pathways by which these germs can penetrate the body and cause infections;
- The correct methods of containment of these germs so that they do not exceed these routes.

2.1 RISK-BASED CLASSIFICATION OF MICROORGANISMS

Microorganisms based on their ability to produce infections, some are harmless; others may be responsible for diseases with mild symptoms; some others can cause serious illnesses; and a small number of them have the capacity to infuse themselves in the community and cause a serious epidemic disease. Experience and research in infections acquired in the laboratory have allowed researchers to classify microorganisms and viruses into three or four groups. Three different systems [3–5] have been published and presented in Tables 2.1 and 2.2, the most recent of them, the Security Measures of the Microbiology Program of the World Health Organization (Table 2.3) [5, 6] is the best and the one that is adopted to this book. This classification uses four Risk Groups (I-IV) in increasing order of danger for those who work in the laboratory and for the community. The Risk Group numbers also indicate the levels of containment, that is, the precautions and techniques necessary to prevent infections. The World Health Organization did not publish relationships of microorganisms within each of the Risk Groups. Member states were urged to prepare their own relationships, since the germs that may present a high risk in one country may be of little or no risk in another; also, it is not necessary to take extreme and costly precautions against a germ in the laboratory if personnel are exposed to it outside. Some relations have been proposed, for example, in the United Kingdom [4] and the US [3, 7], but there is still no general agreement within these countries and between them. Tables 2.1 and 2.2 have classified the germs mentioned in this book within groups I, II, and III, regarding existing lists [4, 7], after requesting the advice of British, American, and European microbiologists who are currently working with germs and after considering the data of the infections that have caused [1, 2]. Group IV is not referred to in this book because it contains only viruses.

2.2 WHO RISK GROUP SYSTEM

2.2.1 RISK GROUP I (LOW INDIVIDUAL AND COMMUNITY RISK)

A microorganism that is unlikely to produce human disease or animal disease or that has veterinary importance.

2.2.2 RISK GROUP II (MODERATE INDIVIDUAL RISK, LIMITED COMMUNITY RISK)

A pathogen that can cause human or animal disease but is unlikely to be a serious risk to laboratory personnel, the community, livestock, or the environment. Laboratory exposures can cause serious infection, but effective treatment and preventive measures are possible, and the risk of diffusion is limited.

2.2.3   *RISK GROUP III (HIGH INDIVIDUAL RISK, LOW COMMUNITY RISK)*

A pathogen that generally causes serious human disease, but that does not usually spread from one infected individual to another.

2.2.4   *RISK GROUP IV (HIGH INDIVIDUAL AND COMMUNITY RISK)*

A pathogen that generally causes serious human or animal disease and can be easily transmitted from one individual to another, directly or indirectly.

Some relationships have been proposed, for example, in United Kingdom [4] and in the US [3, 7], but there is not yet a general agreement within these countries and between them. In Tables 2.1 and 2.2 the germs mentioned in this book have been classified into groups I, II, and III, with reference to existing lists [4, 7], after requesting the advice of British, American, and European microbiologists who are currently working with germs and after considering the data of the infections they have caused [1, 2]. There is no reference in this book to group IV because it contains only viruses.

2.3   INFECTION PATHWAYS

Microorganisms can penetrate the body through the mouth (ingestion), the lungs (inhalation), the skin (injection) and eyes. They can be ingested when pipetting with the mouth and can also penetrate by the fingers and contaminated items on laboratory tables, for example, cigarettes, food, pencils, etc. Such environmental contamination may result from spills and splashes that are not discovered or that are improperly disinfected.

TABLE 2.1　Risk Groups of Fungi and Yeasts Mentioned in This Book

Organism	Risk Group	Organism	Risk Group
Acremenium	I	*Kloeckera*	I
Alternaria	I	*Madurella*	II
Aspergillus	I[a]	*Microsporon*	II
Blastomyces	II	*Neurospora*	I
Botrytis	I	*Paecilomyces*	I
Candida	I	*Paracoccidioides*	II
Cladosporium	I	*braziliensis*	I
Coccidioides immitis	III	*Penicillium*	I
Cryptococcus neoformans	II	*Phialophora*	I
Debaromyces	I	*Pichia*	I
Endomyces	I	*Pullularia*	I
Epidermophyton	III	*Rhodotorula*	I
Fonseceae	I	*Saccharomyces*	I
Geotrichum	I	*Scopulariopsis*	I
Gliocladium	I	*Sporobolomyces*	II
Hansenula	I	*Sporothrix*	I
Helminthosporium	I	*Torulopsis*	II
Histoplasma capsulatum	III	*Trichoderma*	II
		Trichophyton	

[a]A safety cabinet is used for experiments that generate spores. They can be allergens.

Inhalation of infected airborne particles (aerosols), which are released during many common laboratory manipulations, have probably caused the greatest number of laboratory-related diseases [1, 2].

Injection may result from accidental wounds with hypodermic needles, Pasteur pipettes or infected broken glassware. Germs can also penetrate the body through cuts and abrasions of the skin, some of them small enough to go unnoticed by the researcher himself. Splashes of infected fluids in the eyes have caused many infections, some fatal.

The infection pathways in the laboratory are not necessarily the same as naturally acquired infections. In addition, the infective dose may be much higher and, therefore, the symptoms may be different.

TABLE 2.2   Bacterial Risk Groups

Organism		Risk Group	Organism		Risk Group
Acetobacter		I	Microbacterium		I
Achromobacter		I	Micrococcus		I
Acinetobacter		I	Moraxella		I
Actinobacillus		I	Mycobacterium		I
Actinomyces	bovis	II		africanum	III
	eriksonii	II		avium	III
	israelii	II		bovis	III
	naeslundii			chelonei	II
	other ssp.	I		fortuitum	II
Aerococcus		I		intracellulare	III
Aeromonas		I		kansasii	III
Alcalescens		I		leprae	II
Alkaligenes		I		marinum	II
Arizona		I		malmoense	III
Bacillus anthracis		II		scrofulaceum	III
Bacteroides		I		simiae	III
	other spp.	I		szulgai	III
Bifidobacterium		I		tuberculosis	III
Bordetella		II		ulcerans	II
Borrelia		II		xenopi	III
Branhamella		I		other spp.	II
Brevibacterium		I	Neisseria	Gonorrhea	II
Brochothrix		I		meningitides	II
Brucella		III		other spp.	I
Campylobacter		I	Nocardia		II
Chromobacterium		I	Pasteurella		II
Citrobacterium		I	Pediococcus		I
Clostridium	botulinum	III	Photobacterium		I
	difficile	II	Photobacterium		I
	fallax	II	Plesiomonas		I
	novyi	II	Propionibacterium		I
	perfringens	II	Proteus		I

TABLE 2.2　*(Continued)*

Organism		Risk Group	Organism		Risk Group
	septicum	II	*Providence*		I
	sordelli	II	*Pseudomonas*	*mallei*	III
	tetani	II		*pseudomallei*	III
	other ssp.	I		other ssp.	I
Corynebacterium	*diphtheriae*	I	*Salmonella*	*paratyphi A*	III[a]
	equi	I		*typhi*	III[a]
	pyogenes	I		other serotypes	II
	renale	I	*Serratia*		I[b]
	ulcerans	II	*Shigella*		II
	other ssp.	I	*Staphylococcus*	*aureus*	II
Edwardsiella		I		other ssp.	I
Eikenella		I	*Streptobacillus*		II
Enterobacter		I	*Streptococcus*	pathogens	
Erwinia		I		humans and animals	II
Erysipelothrix		III		milk spp.	I
Escherichia		I	*Streptomyces*	*madurae*	II
Flavobacterium	meningo-septicum	I		*pelleteri*	II
	septicum	I		*somaliensis*	II
	other ssp.	I	*Treponema*		II
Francisella		III	*Veillonella*		I
Fusobacterium		I	*Vibrio*	*cholera*	II
Gardnerella		I		parahaemo-lyticus	I
Gemella		I		other spp.	I
Gluconobacter		I	*Yersinia*	*pestis*	III
Klebsiella		I		other spp.	I
Lactobacillus		I			
Legionella		II			
Leptospira		II			

TABLE 2.2 *(Continued)*

Organism	Risk Group	Organism	Risk Group
Leuconostoc	I		
Listeria	II		

[a]Requires a safety laboratory but not a safety cabinet; airborne infections are unlikely.
[b]A safety booth is used for aerosol experiments.

2.4 AEROSOLS: AIRBORNE INFECTED PARTICLES

It is likely that these particles are released and inhaled or contaminate hands and benches, during the following operations: work with handles and syringes, pipetting, centrifugation, crushing, and homogenization, pouring, opening of tubes and culture plates.

2.5 PREVENTION OF INFECTIONS ACQUIRED IN THE LABORATORY: A PRACTICAL CODE

The contention principles involve the creation of:

1. **Primary Barriers:** These around microorganisms to prevent their dispersion in the laboratory;
2. **Secondary Barriers:** These around the operator to act as a safety net if the primary barriers are broken;
3. **Tertiary Barriers:** These prevent the community from reaching any germ that is not contained by the primary and secondary barriers.

This practical code is based mainly on the needs and recommendations set forth in other publications [2–4, 7–9].

2.5.1 PRIMARY BARRIERS

They are the techniques and equipment designed for the containment of microorganisms and to prevent them from accessing the operator directly and spreading them as aerosols:

- Pipetting with the mouth must be prohibited in all cases. It is necessary to provide mechanical devices for pipetting.
- No objects should be brought to the mouth. Glass pieces for rubber tubes attached to pipettes, pens, pencils, labels, smoking utensils, fingers, food, and drink must be included among them.
- The use of hypodermic needles should be restricted. Cannulas are safer.
- Pointed glass Pasteur pipettes should be replaced by soft plastic ones.
- Cracked and chipped glass material must be replaced.
- Centrifuge tubes should be filled only up to 2 cm from the edge. All Risk Group III products must be centrifuged in tightly closed centrifuge tubes.
- Bacteriological inoculating loops must be fully closed, with a diameter not exceeding 3 mm and with a stem no larger than 5 cm and must be held in metal [2] handles, not glass.
- Homogenizers should be inspected regularly to discover defects that may disperse aerosols. Only maximum safety models should be used [2]. Griffith glass tubes and tissue homogenizers should be held with a wadding pad and hand protected by a glove when working.
- All the products of Risk Group III must be processed in a Safety Laboratory and in a Microbiological Safety Booth unless stated in Table 2.1.
- A provision of disinfectant must be available at each work point.
- Benches and work surfaces must be disinfected regularly and after spillage of contaminated material.
- At each workstation there must be waste containers for small objects and reusable pipettes. They must be emptied, decontaminated, and filled with freshly prepared disinfectant daily.
- Disposable containers or bags held within containers must be placed in the vicinity of each workstation. Autoclave should be removed and sterilized daily.
- Broken culture vessels, for example, after an accident, should be covered with a cloth. A suitable disinfectant must be spilled on the cloth and left there for 30 minutes. The remains should be removed in a suitable container (tray or dustpan) and sterilized in

the autoclave. Work with gloves and collect the remains with hard cardboard.

- Infectious material that must be sent by mail or by air must be packaged in accordance with government, postal or airline regulations, which can be obtained from those authorities. In other places full instructions are also given [2, 4, 6, 8].
- Sample containers must be strong and airtight.

2.5.2 SECONDARY BARRIERS

These measures are aimed at protecting personnel if primary barriers fail. However, they should be observed as strictly as the previous ones:

- Lab suits and coats should always be worn and fastened. They should be kept separate from street dresses and other clothes.
- Protective laboratory clothing should be removed when personnel leave the laboratory and not wear it in other areas.
- Hands should be washed after handling infectious material and always before leaving the laboratory.
- Visible cuts, scrapes, and abrasions on exposed parts of the body of the personnel should be coated with waterproof dressings.
- There should be medical supervision in laboratories where pathogenic germs are handled.
- Any illness should be reported to the doctor or occupational health supervisors. It can be related to the laboratory. Pregnancy must also be manifested. It is inadvisable to work with certain germs during it.
- Any staff member receiving steroids or other immunosuppressive medications should tell the medical supervisor.
- When possible and achievable, staff should be vaccinated against probable infections [2, 4].
- In laboratories working with tuberculosis, personnel must have received BCG vaccine or have shown a positive skin reaction before starting work. Subsequently they must take X-ray exams annually [2, 4].

2.5.3   *TERTIARY BARRIERS*

They are designed to offer additional protection to personnel and to prevent the dissemination of the microorganisms under investigation in the laboratory. They comprise architectural and engineering design and the reader of this book should consult other publications [2, 6, 10]. The WHO risk group system requires three levels of laboratory design, which are summarized in subsections.

2.5.3.1   BASIC LABORATORY

It is designed to work with germs from Risk Groups I and II, it must have plenty space. The walls, ceilings, and floors must be smooth, non-absorbent, easy to clean and disinfect, and must be resistant to chemical agents. Also, the floors must be slip proof, the lighting and heating must be adequate. Basins other than laboratory batteries are essential.

The upper part of the benches should be wide, at the correct height to work comfortably in a sitting position, smooth, easy to clean and disinfect and resistant to the chemical agents that will probably be used. They must have adequate storage possibilities. Access must be limited to authorized persons.

2.5.3.2   SAFETY LABORATORY

It is designed to work with germs from Risk Group III. They must have incorporated all the requirements indicated for the Basic Laboratory and must be physically separated from other rooms, without communication (for example, pipes, suspended ceilings) with other areas, apart from the door, which must be locked and transferable grilles for ventilation. Ventilation must be in one direction, achieved by a lower pressure than in the other rooms and adjacent areas. The air must be removed to the atmosphere (total discharge, without recirculation to other parts of the building) with an extraction system coupled to the microbiological safety cabin so that during working hours the air is continuously extracted and the airborne particles cannot move from one side of the building to the other. Replacement air penetrates through the transferable grilles. Access must be strictly regulated. The international biohazard warning sign must be posted on the door, with the appropriate expression (Figure 2.1).

FIGURE 2.1   International biological risk symbol.

2.5.3.3   MAXIMUM SECURITY LABORATORY

It is required to work with Risk Group IV products and is beyond the scope of this book. Its construction and use generally require licensing and supervision of the authorities.

2.6   EDUCATION IN SAFETY WORK

Education must be part of the general training that all microbiological laboratory personnel should receive and should be included in Good Laboratory Practices. In general, the methods that protect microbial growth from contamination also protect against infection, although this should not be taken for granted. Personnel protection can only be obtained through good training and careful work. Programs for safety training in microbiology have been published.

TABLE 2.3 Summary of the Risk-based Classification of Microorganisms for Laboratory Personnel and for the Community

	Low Risk			**High Risk**
USPHS (1974)	Class 1	Class 2	Class 3	Class 4
	None or minimum	Ordinary potential	Special for a person	High, for a person and the community
United Kingdom (1978)[a]	–	Category C	Category B	Class A
		No special potential	Special for a person	High, for a person and the community
WHO (1979)	Risk group I individual, low community, low	Risk Group II individual, moderate community, limited	Risk Group III individual, high community, low	Risk Group IV Individual, high community, high
Appropriate laboratory	Basic	Basic	Security[a]	Maximum security[b]

[a]DHSS (1978).
[b]Refers to precautions, equipment, and laboratory design.

KEYWORDS

- **aerosols**
- **microbiological laboratory personnel**
- **microbiology**
- **microorganisms**
- **pathogen**
- **World Health Organization**

REFERENCES

1. Pike, R. M., (1979). Laboratory-associated infections: Incidence, fatalities, causes, and prevention. *Anu. Rev. Microbiol., 33*(1), 41–66.
2. Cook, D. A., (2000). Laboratory-acquired infections: History, incidence, causes and preventions. *J. Antimicrob. Chemoth., 45*(6), 933.

3. *Classification of Etiologic Agents on the Basis of Hazard* (4th edn.). Atlanta, Ga.: Center for Disease Control, Office of Biosafety, 19741975.
4. Howie, J. W., & Collins, C. H., (1980). The Howie code for preventing infection in clinical laboratories: Comments on some general criticisms and specific complaints. *The British Medical Journal, 280*(6221), 1071–1074.
5. World Health Organization, (1979). *Weekly Epidemiological Record, 54*, 44.
6. World Health Organization, (2004). *Laboratory Biosafety Manual, 5–170.*
7. Collins, C. H., (1990). Safety in industrial microbiology and biotechnology: UK and European classifications of microorganisms and laboratories. *Trends Biotechnol., 8*, 345–348.
8. US Department of Health and Human Services, (2004). *Biosafety in Microbiological and Biomedical Laboratories* (pp. 1–405). Centers for Disease Control and Prevention National Institutes of Health.
9. Richmond, M., (1991). *The Microbiological Safety of Food, Part II, 220.*
10. Green, W. E., & Donaldson, D., (1990). Safety measures to be taken when moving to a new laboratory. In: Pal, S. B., (ed.), *Handbook of Laboratory Health and Safety Measures*. Springer, Dordrecht.

CHAPTER 3

Food-Microorganism Interaction

NATHIELY RAMIREZ-GUZMAN,[1] LEONARDO SEPÚLVEDA,[2] and
CRISTÓBAL NOÉ AGUILAR[2]

*[1]Center for Interdisciplinary Studies and Research,
Autonomous University of Coahuila, Saltillo, México,
E-mail: nathiely.ramirez@uadec.edu.mx*

*[2]Bioprocesses and Bioproducts Research Group, Food Research
Department, Autonomous University of Coahuila, Saltillo, México*

3.1 INTRODUCTION

The food sector has always suffered great losses due to the sensitivity and instability of its products, which will depend on their composition and nature, their main threat being microorganisms [1]. All foods have an endogenous microflora, extremely variable, concentrated mainly in the superficies, and in some internal tissues, they eventually have viable microbial forms, whose development depends on the Intrinsic Factors [2]. Besides the endogenous or food-based microflora, they are subject to contamination by different microorganisms because of the stages they undergo during their processing, handling, contact with equipment, utensils, or due to the environmental conditions under which they are processed [3]. These are known as extensive factors. The predominant species or groups of microorganisms in food depend strongly on the characteristics inherent to those foods (intrinsic factors), and on the unit operations to which they undergo during their transformation or conservation (extrinsic factors) [4].

Bacteria and yeasts are the most important microorganisms as potential agents of deterioration or as possible pathogens for man. In most cases, bacteria are the predominant microorganisms in food, mainly because [5]. They have a very short generation time, they can use various substrates, present a wide variety of behaviors (various genres) in response to environmental conditions [6]. The knowledge of the intrinsic or extrinsic factors that favor or inhibit the multiplication of microorganisms is essential to understand the basic principles that govern both the alteration and food conservation [4].

3.2   MICROBIAL DEVELOPMENT INTRINSIC FACTORS

3.2.1   WATER ACTIVITY (AW)

Metabolism and the growth of the microorganisms require the presence of water in an available form, and Aw is an index of its availability for use in chemical reactions and microbial growth. Aw is by definition, the ratio between the vapor pressure of a solution over the vapor pressure of pure water, or it can also be estimated as the number of moles of the solute between the sum of the number of moles of the solute plus the number of moles of the solvent [5]. So, it is easily predictable that if solutes are added to a pure solvent, there will be a reduction in the vapor pressure of the solution and as a consequence the value of Aw will decrease, with 1 being the value of Aw for pure water, for what the Aw values are within a range of 0 to 1 [7]. For example, a medium with 0.88% NaCl, 8.52% sucrose and 4.45% glucose has a value of Aw = 0.995, while a medium with 18.18% NaCl, 68.60% sucrose and 58.45% glucose has a value Aw = 0.86.
Eqn. (1), presents the equation that defines Aw in terms of vapor pressure and the molar relationship between solute and solvent.

$$Aw = \frac{P}{P_0} = \frac{n_2}{n_1 + n_2} \qquad (1)$$

where; P is the vapor pressure of the solution; Po is the vapor pressure of the pure solvent; n_1 is the number moles of solute; and n_2 is the number of moles of solvent [8].

The factors in food capable of reducing water vapor pressure and consequently Aw are: Adsorption of water molecules on the surface, Capillary forces, Formation of solutions with different solutes, Formation of hydrophilic colloids, Presence of crystallization or hydration water. For example, glycerol is an Aw depressant and is generally used in culture media to inhibit the growth of microorganisms very sensitive to Aw changes, such as bacteria, since most of them do not grow at lower values of 0.91. Table 3.1 presents the minimum Aw values that allow the multiplication of microorganisms that alter food [7, 9].

TABLE 3.1 Aw Values Necessary for Microbial Growth

Microorganism Group	Aw Minimum Value
Most bacteria	0.91
Most yeast	0.88
Most fungi	0.80
Halophilic bacteria	0.75
Xerophilous fungi	0.65
Osmophilic yeast	0.60

The Aw values in which the different types of microorganisms can subsist and reproduce means that foods that due to their composition and nature are between these Aw levels make them susceptible to attack by these pathogens, in this chapter are foods, which makes them more susceptible to different microorganisms according to the values of water activity (WA) they possess, for which five groups are classified in this chapter according to the range of Aw value they possess, as observed in Table 3.2.

TABLE 3.2 The Main Food Groups and Their Aw Values

Food Groups	Aw Values
Fresh meat and fish, fresh fruits and vegetables, milk, and most drinks, canned vegetables in brine, canned fruits with low sugar concentration	0.98 and above
Tomato paste, cheese subjected to industrial treatment, cured, and canned meats, sausages, fermented (not dehydrated), and canned fruits with high sugar concentration.	0.93 to 0.98
Dry and fermented sausages, cheddar cheeses and condensed milk	0.85 to 0.93
Dried fruits, flour, cereals, compotes, and jellies, nuts, and intermediate moisture foods.	0.80 to 0.85
Chocolate, pastry, honey, biscuits, snacks, English potatoes, eggs with vegetables, powdered milk, and cookies.	0.60 to 0.85

3.2.2 HYDROGEN POTENTIAL (PH)

The pH of a system is obtained through the relationship pH = –log [H⁺].
When the concentration of hydrogen ions is higher in the system, the lower
its pH value will be. Depending on the pH values of the food, these can be
classified as [10]:

1. **Low Acid Foods (pH> 4.5):** Under these conditions, there is a
   predominance of microbial growth, which can occur in better
   generation times, so they may be susceptible to attack by pathogenic, sporogenic, aerobic, anaerobic, mesophilic, and thermophilic microorganisms [11].
2. **Acidic Foods (pH between 4.5 and 4.0):** Under these conditions,
   microbial growth is restricted to oxidative or fermentative yeasts
   and filamentous fungi. Sporogenic and non-sporogenic bacteria
   can also develop [12].
3. **Very Acidic Foods (pH <4.0):** In these foods grow exclusively
   yeasts and fungi, acetic bacteria, and *Zymomonas* that develop at
   pH's close to 3.7 [13].

It is important to note that acids can exert an inhibitory or lethal effect
on microbial cells, due to the high concentration of hydrogen ions or by
the same toxicity of undissociated acids, also, the external pH conditions
in the culture medium seriously affect the intracellular pH values [14].
Acidification inside the cell can occur by migration from hydronium
ions of the external environment, by the dissociation of molecules that
penetrate through the membrane and form acids, which interfere with
the same permeability, seriously affecting nutrient transport systems
and oxidative phosphorylation, inhibiting the transport of electrons and
causing acidification inside the cell, and also because some acids when
dissociated release anions that can be metabolized generating intermedi-
ates with inhibitory activity [15, 16].

3.2.3 OXIDE-REDUCTION POTENTIAL, REDOX (EH)

Redox potential is an intrinsic factor highly dependent on the extrinsic
factors of microbial development, facilitating food alteration or the
susceptibility to chemical modifications caused by microorganisms.

These alterations at the end of the day are summarized in how easy electrons are donated or accepted between the different chemical compounds present in the systems (food) in question. It is important to remember that the release or loss of electrons is considered as the oxidation process, while the reception of electrons is known as reduction [17, 18].

The relationship between both processes, oxidation, and reduction, is established through the Nernst equation (Eh):

$$Eh = Eo + 0.06/n \log (ox)/(red) \tag{2}$$

where; Eh is the redox potential (mV); Eo is the standard potential (pH = 0); n is the number of electrons involved in the process; (ox)/(red) is the concentration in the oxidized and reduced states.

The lower the value of Eh, the greater the ability to transfer electrons. The redox potential is determined with potentiometers. Aerobic microorganisms have Eh growth values of +350 to +500 mV, anaerobes grow at Eh values between +30 to –250 mV, but preferably at values of +150 mV. Finally, optional anaerobes grow in Eh ranges from + 100 to +350 mV.

The presence of molecular oxygen for anaerobes is more lethal than for the rest of the groups, since this compound makes the redox potential positive being the last electron acceptor compound. This behavior is explained by the fact that microorganisms are unable to express the catalase activity that breaks down the hydrogen peroxide, the accumulating in the cell and creating a toxic environment [19].

3.3 MICROBIAL DEVELOPMENT EXTRINSIC FACTORS

3.3.1 TEMPERATURE

Temperature is extremely important since it intervenes with the metabolism of the cell, microbial growth occurs in a wide range of temperatures, ranging from –8 to 90°C. The temperature has a very strong influence on: Lag phase duration, Growth speed, The final number of cells in a population, Chemical, and enzymatic composition of the cell [20]. Depending on the optimal temperature and the growth of the microorganisms, these can be grouped as in Table 3.3.

TABLE 3.3 Types of Microorganisms According to Their Optimal Temperature

Microorganisms	Temperature (°C)
Thermophiles	55–75
Mesophiles	30–45
Psychrotrophiles	25–30
Psychrophiles	12–15

The environment characterizes the classification according to their optimal development temperature and particular characteristics of microorganisms, for example, Psychrotrophiles are extremely important as agents that alter food, especially those that belong to the genera of *Pseudomonas, Actinobacter, Vibrio, Lactobacillus*, and *Bacillus*. Most of the thermophilic bacteria of food importance are included in the genera *Bacillus* and *Clostridium*, but few species of these genera are considered thermophilic. Among mesophiles stand out the pathogenic bacteria and detractors of food quality, some fungi, and yeasts. Among the strict psychrophiles are marine microorganisms and those that grow in environments whose temperature remains constantly reduced.

3.3.2   RELATIVE HUMIDITY OF THE ENVIRONMENT-RH

Perishability is directly associated with food, due to the close relationship between RH and Aw, given by the relationship:

$$\% \, RH = Aw \times 100$$

For a better or adequate packaging of food products and in order to guarantee or control microbial development, lengthening the shelf life of them, the Aw/RH must be considered [21].

3.3.3   GASEOUS COMPOSITION OF THE ENVIRONMENT

The uses of controlled atmospheres can modify the processes of food spoilage by accelerating or delaying it, for example, the presence of oxygen slows the spoilage of food by anaerobic microorganisms. Carbon dioxide is the gas most used to slow down the deterioration of food (fruits and meat products). Most yeasts, bacteria, and fungi are inhibited but not destroyed in atmospheres containing between 5 and 50% CO_2. This gas is very effective against psychrotrophilic microorganisms, Gram negative bacteria, fungi, and oxidative yeasts. On those who tolerate its presence, it doubles the time of its multiplication. Gram positive bacteria, mainly lactic ones, are resistant to high concentrations of CO_2 [22, 23].

3.4 CHEMICAL MODIFICATIONS CAUSED BY MICROORGANISMS

3.4.1 MODIFICATIONS OF NITROGEN COMPOUNDS

Most of the nitrogen contained in food is part of the proteins, which can be used as sources of nitrogen by microorganisms given their ability to produce proteases, or by the same action of endogenous proteases that lead proteins to peptides and amino acids. Some peptides formed in aerobiosis have a bitter taste but not unpleasant, which modifies the food taste and is an index to establish the alteration of them. If peptides and amino acids are formed in anaerobiosis, they impart unpleasant and rotten flavors and odors to food, because most of them have sulfur compounds, such as methyl, and ethyl mercaptan, amines, hydrogen sulfides, and ammonia [24, 25]. During microbial growth, microorganisms can use different ways of using amino acids, such as Table 3.4.

TABLE 3.4   Use of Amino Acids by Microorganisms

Chemical Reaction	Products
Oxidative deamination	Keto acids + NH_3
Hydrolytic deamination	Hydroxy acid + NH_3
Reductive deamination	Saturated fatty acid + NH_3
Denaturalization o deamination	Non-saturated fatty acid + NH_3
Oxidation-reduction	Keto acid + fatty acid + NH_3
Decarboxylation	Amide + CO_2
Hydrolytic deamination + Decarboxylation	Primary alcohol + NH_3 + CO_2
Reductive deamination + Decarboxylation	Hydrocarbon + NH_3 + CO_2
Oxidative deamination + Decarboxylation	Fatty acid + NH_3 + CO_2

3.4.2   CARBOHYDRATE MODIFICATIONS

Carbohydrates are the substrates preferred by microorganisms and are one of the most abundant compounds in food. Poly, oligo, and disaccharides must be enzymatically hydrolyzed to simpler compounds, monosaccharides [26].

Glucose is one of the substrates of easier microbial assimilation when used under Aerobiosis, oxidizes to carbon dioxide and water or Anaerobiosis, it is incompletely oxidized to different compounds that can be formed depending on the fermentation pathway for example:

1. **Alcoholic Fermentation:** Yeasts produce alcohol and CO_2.
2. **Lactic Fermentation:** Homofermentative lactic bacteria generates only lactic acid.
3. **Mixed Lactic Fermentation:** Heterofermentative lactic bacteria produce lactic acid, acetic acid, ethanol, glycerol, and water.
4. **Coniform Fermentation:** Coniform bacteria produce lactic, acetic, formic acid, ethanol, acetoin, butanediol, hydrogen, and carbon dioxide.
5. **Propionic Fermentation:** Propionic bacteria produce propionic, succinic, and acetic acid besides carbon dioxide.
6. **Butyric-Butyl-Iso-Propyl Fermentation:** Anaerobic bacteria produce butyric acid, acetic acid, hydrogen, acetone, butylene glycol, butanol, 2-propanol, and carbon dioxide.

Additionally, it is important to note that some microorganisms can use carbohydrates to form a variety of compounds that include higher fatty acids, organic acids, aldehydes, ketones, and esters [27, 28].

3.5   IMPORTANT MICROORGANISMS IN FOOD MICROBIOLOGY

In food microbiology, microorganisms are normally classified into three groups: indicator microorganisms that correspond to those bacteria, fungi or yeasts that cause damage in the different stages of the production chains starting with the cultivation, processing, transport, and storage, being responsible for large losses for the food sector, pathogenic microorganisms agents that cause diseases in humans such as poisonings and infections that cause hospitalizations, and the initiators used for the manufacture of products of industrial interest such as food or food additives [4, 29].

3.5.1 PATHOGENIC MICROORGANISMS

Zoonotic diseases constitute a group of animal diseases that are transmitted to man by direct contagion with the sick animal, through somebody fluid, or the presence of an intermediary such as mosquitoes or other insects. They can also be contracted by consumption of foods of animal origin that do not have the corresponding sanitary controls, or by consumption of poorly washed raw fruits and vegetables. However, microbial agents such as parasites, viruses, or bacteria can also cause them [30]. Table 3.5 shows some examples.

TABLE 3.5   Zoonotic Diseases

Zoonotic Diseases	Causative Agent
Anthrax	*Bacillus anthracis*
Tuberculosis	*Mycobacterium tuberculosis*
Listeriosis	*Listeria monocytogenes*
Brucellosis	*Brucella abortus, B. melitensis*
Tuleramia	*Francisella tulharensis*
Query fever	*Coxiella burnetti*

For its part, food poisoning is produced by the ingestion of a microbial product called toxin that the bacteria produce under stress, the main examples are Staphylococcal poisoning caused by *Staphylococcus aureus*, a common bacterium in the environment and surfaces, and Botulism produced by the causative agent *Clostridium botulinum* related to packaged foods [31].

Food infections are the most common and recurrent associated with poor eating practices, as well as poor hygiene habits, they can range from a simple stomachache to more complicated diseases that can lead to death. the main agents and diseases they cause are Enterobacteria: *Shigella* (shigellosis); *Salmonella* (salmonellosis, typhoid fever, paratyphoid fever); *Yersinia enterocolitica* (yersiniosis); *Escherichia coli*. Bacteria of importance in public health: *Campylobacter*; *Vibrio cholerae*; *Vibrio parahaemolyticus*. Gram-positive sporulated pathogenic bacteria: *Clostridium perfringes*; *Bacillus cereus* (gastroenteritis), etc. [32].

Finally, as less common but no less important are the mycotoxins that can contaminate the food chain because of the infection of agricultural products intended for human consumption or domestic animals. The complication with these is their resistance to decomposition and destruction during digestion, which is why they remain in the food chain and in dairy products. They even resist cooking and freezing. The most common toxins in agricultural products are produced by species of the genera *Aspergillus*, *Penicillium*, and *Fusarium*, among others [11].

3.5.2   MICROORGANISMS USED IN THE PRODUCTION OF FOOD, ADDITIVES, AND ENZYMES

At present in different industries in addition to food, such as cosmetics, textiles, pharmaceuticals, paper, among others, one of its main allies for its productions is the implementation of biotechnology using microorganisms, since for years they have proven to be an important source of production of different compounds, as a result of the enormous metabolic machinery they possess, humanity has learned to manipulate and make the most of it, some of the examples are [33–35]: ethanol production yeasts; organic acid production: vinegar production acetic bacteria – *Acetobacter*; lactic acid production – lactic bacteria; propionic acid production – Propionic bacteria; citric acid production; gluconic acid production; fumaric acid production; gibberellin acid production [13, 36], unicellular protein production, enzyme production as lipases, cellulases, tanases, proteases, amylases, among others, they are of utmost importance in processes such as clarification of juices, elaboration of cleaning products, paper production, as well as the formation of various drugs [37–39].

   One of the main and with which the biotechnological revolution began is for the production of antibiotics such as penicillin, one of the most important events for the history of medicine and the whole of humanity since the discovery of a substance, a metabolite that produced a fungus came to revolutionize the treatment of infections [28]. And of course, and we cannot exclude the use of microorganisms in food technology, transforming, and creating food from ancient times with the production of wine and bread, through conventional ones, cheeses, yogurt, beer, vinegars, etc., even in the present with the new trends in the production of functional fermented foods [40].

KEYWORDS

- *Bacillus*
- *Lactobacillus*
- microflora
- microorganisms
- water activity
- *Zymomonas*

REFERENCES

1. Ramírez-Guzmán, N., Torres-León, C., Martínez-Terrazas, E., De La Cruz-Quiroz, R., Flores-Gallegos, A. C., Rodríguez-Herrera, R., & Aguilar, C. N., (2018). Biocontrol as an efficient tool for food control and biosecurity. In: Grumezescu, A. M., & Holban, A. M., (eds.), *Food Safety and Preservation* (pp. 167–193). Elsevier: United Kingdom.

2. Lin, P. Y., Wood, W., & Monterosso, J., (2015). Healthy eating habits protect against temptations. *Appetite, 103*, 432–440.

3. Smet, C., Baka, M., Steen, L., Fraeye, I., Walsh, J. L., Valdramidis, V. P., & Van, I. J. F., (2019). Combined effect of cold atmospheric plasma, intrinsic and extrinsic factors on the microbial behavior in/on (food) model systems during storage. *Innov. Food Sci. Emerg. Technol., 53*, 3–17.

4. Smet, C., Noriega, E., Rosier, F., Walsh, J. L., Valdramidis, V. P., & Van, I. J. F., (2016). Influence of food intrinsic factors on the inactivation efficacy of cold atmospheric plasma: Impact of osmotic stress, suboptimal PH and food structure. *Innov. Food Sci. Emerg. Technol., 38*, 393–406.

5. Ozogul, Y., Kuley, B. E., Akyol, I., Durmus, M., Ucar, Y., Regenstein, J. M., & Köşker, A. R., (2020). Antimicrobial activity of thyme essential oil nanoemulsions on spoilage bacteria of fish and food-borne pathogens. *Food Biosci., 36*, 100635.

6. Riu, J., & Giussani, B., (2020). Electrochemical biosensors for the detection of pathogenic bacteria in food. *TrAC - Trends Anal. Chem., 126*, 115863.

7. Subbiah, B., Blank, U. K. M., & Morison, K. R., (2020). A review, analysis and extension of water activity data of sugars and model honey solutions. *Food Chem., 326*, 126981.

8. Placido, M., & Alemán, M., (2002). Hygrometric Method for Determining Water Activity. *CYTA – J. Food., 3*(4), 229235.

9. Yang, R., Guan, J., Sun, S., Sablani, S. S., & Tang, J., (2020). Understanding water activity change in oil with temperature. *Curr. Res. Food Sci., 3*, 158–165.

10. Téllez-Luis, S. J., Ramírez, J. A., Pérez-Lamela, C., Vázquez, M., & Simal-Gándara, J., (2001). Application of high hydrostatic pressure in the food preservation. *CYTA – J. Food 3*(2), 66–80.

11. Huang, Y., Wilson, M., Chapman, B., & Hocking, A. D., (2010). Evaluation of the efficacy of four weak acids as antifungal preservatives in low-acid intermediate moisture model food systems. *Food Microbiol., 27* (1), 33–36.

12. Hernández-Almanza, A., Montañez, J., Martínez, G., Aguilar-Jiménez, A., Contreras-Esquivel, J. C., & Aguilar, C. N., (2016). Lycopene: Progress in microbial production. *Trends Food Sci. Technol., 56*, 142–148.

13. Mohd, Y. N. H., Rahman, N. A., Man, H. C., Mohd, Y. M. Z., & Hassan, M. A., (2011). Microbial characterization of hydrogen-producing bacteria in fermented food waste at different PH values. *Int. J. Hydrogen Energy, 36*(16), 9571–9580.

14. Blandino, A., Al-aseeri, M. E., Pandiella, S. S., Cantero, D., & Webb, C., (2003). Cereal-based fermented foods and beverages. *Food Res. Int., 36*, 527–543.

15. Tokpah, D. P., Li, H., Wang, L., Liu, X., Mulbah, Q. S., & Liu, H., (2016). An assessment system for screening effective bacteria as biological control agents against *Magnaporthe grisea* on rice. *Biol. Control, 103*, 21–29.

16. Jiménez-Delgadillo, R., Valdés-Rodríguez, S. E., Olalde-Portugal, V., Abraham-Juárez, R., & García-Hernández, J. L., (2018). Effect of pH and temperature on the growth and antagonistic activity of *Bacillus subtilis* on *Rhizoctonia solani*. *Mex J. Phytopathol.*, *36*(2), 256–275.

17. Franco, R., & Martínez-Pinilla, E., (2017). Chemical rules on the assessment of antioxidant potential in food and food additives aimed at reducing oxidative stress and neurodegeneration. *Food Chem., 235*, 318–323.

18. Sahraee, S., Milani, J. M., Regenstein, J. M., & Kafil, H. S., (2019). Protection of foods against oxidative deterioration using edible films and coatings: A review. *Food Biosci., 32*, 100451.

19. Takumi, S., Komatsu, M., Aoyama, K., Watanabe, K., & Takeuchi, T., (2008). oxygen induces mutation in a strict anaerobe, *Prevotella melaninogenica. Free Radic. Biol. Med., 44*(10), 1857–1862.

20. Rifna, E. J., Singh, S. K., Chakraborty, S., & Dwivedi, M., (2019). Effect of thermal and non-thermal techniques for microbial safety in food powder: Recent advances. *Food Res. Int., 126*, 108654.

21. Ramírez-Guzmán, N., Chávez-González, M., Sepúlveda-Torre, L., Torres-León, C., Cintra, A., Angulo-López, J., Martínez-Hernández, J. L., & Aguilar, C. N., (2020). Significant advances in biopesticide production: Strategies for high-density bio-inoculant cultivation. In: Singh, J. S., & Vimal, S. J., (eds.), *Microbial Services in Restoration Ecology* (pp. 1–11). Elsevier: The Netherlands.

22. McEvoy, P. B., (2018). Theoretical contributions to biological control success. *BioControl, 63*(1), 87–103.

23. Thrane, M., Paulsen, P. V., Orcutt, M. W., & Krieger, T. M., (2017). Soy protein: Impacts, production, and applications. In: Nadathur, S. R., Wanasundara, J. P. D., & Scanlin, L., (eds.), *Sustainable Protein Sources* (pp. 23–45). Academic Press: United Kingdom.

24. Li, M., Ma, G. S., Lian, H., Su, X. L., Tian, Y., Huang, W. K., Mei, J., & Jiang, X. L., (2019). The effects of *Trichoderma* on preventing cucumber fusarium wilt and regulating cucumber physiology. *J. Integr. Agric., 18*(3), 607–617.

25. Liao, Y., Zeng, M., Wu, Z., Chen, H., Wang, H. N., Wu, Q., Shan, Z., & Han, X., (2012). Improving phytase enzyme activity in a recombinant PhyA mutant phytase

from *Aspergillus niger* N25 by error-prone PCR. *Appl. Biochem. Biotechnol., 166*(3), 549–562.

26. Li, X., Xie, X., Xing, F., Xu, L., Zhang, J., & Wang, Z., (2019). Glucose oxidase as a control agent against the fungal pathogen *Botrytis cinerea* in postharvest strawberry. *Food Control, 105*, 277–284.

27. Chávez-García, M., Montaña-Lara, J. S., Martínez-Salgado, M. M., Mercado-Reyes, M., Rodríguez, M. X., & Quevedo-Hidalgo, B., (2009). Effects of substrate and light exposition in the production of *Trichoderma* Sp. *Univ. Sci., 13*(3), 245–251.

28. Ramírez-Guzmán, K. N., Torres-León, C., Martinez-Medina, G. A., De La Rosa, O., Hernández-Almanza, A., Alvarez-Perez, O. B., Araujo, R., et al., (2019). Traditional fermented beverages in Mexico. In: Grumezescu, A. M., & Holban, A. M., (eds.), *Fermented Beverages* (pp. 605–635). Woodhead Publishing: United Kingdom.

29. Monteiro De, S. P., & De Oliveirae, M. P., (2010). Application of microbial α-amylase in industry: A review. *Braz. J. Microbiol., 41*, 850–861.

30. Standley, C. J., Carlin, E. P., Sorrell, E. M., Barry, A. M., Bile, E., Diakite, A. S., Keita, M. S., et al., (2019). Assessing health systems in guinea for prevention and control of priority zoonotic diseases: A one health approach. *One Heal., 7*, 100093.

31. Fusco, V., Besten, H. M. D., Logrieco, A. F., Rodriguez, F. P., Skandamis, P. N., Stessl, B., & Teixeira, P., (2015). Food safety aspects on ethnic foods: Toxicological and microbial risks. *Curr. Opin. Food Sci., 6*, 24–32.

32. Ng, K. R., Lyu, X., Mark, R., & Chen, W. N., (2019). Antimicrobial and antioxidant activities of phenolic metabolites from flavonoid-producing yeast: Potential as natural food preservatives. *Food Chem., 270*, 123–129.

33. Gellen, L. F. A., Silva, D. P., Souza, S. R., Freitas, A. C., Scheidt, G. N., & Nascimento, I. R., (2014). Potencial farmacoindustrial de *trichoderma harzianum* para fins farmacoterapêuticos. *Biota. Amaz., 4*(4), 91–96.

34. De La Cruz-Quiroz, R., Roussos, S., Tranier, M. T., Rodríguez-Herrera, R., Ramírez-Guzmán, N., & Aguilar, C. N., (2019). Fungal spores production by solid-state fermentation under hydric stress condition. *J. BioProcess Chem. Technol., 11*(21), 7–12.

35. Hernández-Almanza, A., Montañez-Sáenz, J., Martínez-Ávila, C., Rodríguez-Herrera, R., & Aguilar, C. N., (2014). Carotenoid production by *Rhodotorula glutinis* YB-252 in solid-state fermentation. *Food Biosci., 7*, 31–36.

36. Seviour, R. J., Harvey, L. M., Fazenda, M., & McNeil, B., (2013). Production of foods and food components by microbial fermentation: An introduction. In: McNeil, B., Archer, D., & Harvey, L., (eds.), *Microbial Production of Food Ingredients, Enzymes and Nutraceuticals* (pp. 97–124). Woodhead Publishing: United Kingdom.

37. Quintero, M. M., & Gutiérrez, S. P. A., (2010). Purification and characterization of a α-amylase produced by *Bacillus* sp. BBM1. *Dyna., 77*(162), 31–38.

38. Kasana, R. C., Salwan, R., & Yadav, S. K., (2011). Microbial proteases: Detection, production, and genetic improvement. *Crit. Rev. Microbiol., 37*(3), 262–276.

39. Torres-León, C., Ramirez-Guzman, N., Ascacio-Valdes, J., Serna-Cock, L., Dos, S. C. M. T., Contreras-Esquivel, J. C., & Aguilar, C. N., (2019). Solid-state fermentation with *Aspergillus niger* to enhance the phenolic contents and antioxidative activity of Mexican mango seed: A promising source of natural antioxidants. *LWT, 112*, 108236.

40. Quintero-Salazar, B., Bernáldez-Camiruaga, A. I., Dublán-García, O., Barrera-García, V. D., & Favila-Cisneros, H. J., (2012). Consumption and knowledge of a traditional fermented drink in Ixtapan del Oro, México: the *sambumbia*. *Alteridades, 22*, 115–129.

CHAPTER 4

Food Preservation

CRISTIAN TORRES-LEÓN[1] and CRISTÓBAL NOÉ AGUILAR[2]

[1]*Research Center and Ethnobiological Garden, Autonomous University of Coahuila, Viesca – 27480, Coahuila, México, E-mail: ctorresleon@uadec.edu.mx*

[2]*Bioprocesses and Bioproducts Research Group, Food Research Department, Autonomous University of Coahuila, Saltillo, México*

4.1 INTRODUCTION

The main objective of food preservation has been on controlling microbial populations, with a specific emphasis on pathogenic microorganisms. Food preservation implies inhibiting the growth of microorganisms [1]. Hostile environments for microorganisms are an adequate food preservation strategy The application of heat treatments, reduction of storage temperatures, application of good manufacturing practices and the addition of additives define the food shelf-life and safety [2]. The most important hurdles are water activity (a_w), temperature (high or low), acidity (pH), dehydration, preservatives, and non-thermal technologies (non-conventional).

4.2 FOOD GROUPS

4.2.1 VEGETABLES

Cereals, sugar, fungi (grown as food), fruits, and vegetables are susceptible to microbial and physical-chemical deterioration after harvest. This generates the need to use preservation techniques. The postharvest preservation foods as of fruits and vegetables is a challenge owing to their highly perishable nature [3].

4.2.2   ANIMALS

Foods like meat and meat products, birds, and eggs, fish (and other marine foods), and milk and derivatives are easily altered by microorganisms unless they undergo some conservation treatment. Most of these foods must be refrigerated or even frozen immediately after harvesting, in order to inhibit microbial growth and loss of quality [1].

4.3   PRINCIPLES OF FOOD PRESERVATION

4.3.1   PREVENTION OR DELAY (SHELF LIFE) OF THE BACTERIAL COMPOSITION

The keeping food without germs (asepsis), the elimination existing germs (filtration), then placing an obstacle to microbial growth (low temperatures, dehydration, anaerobic conditions, chemical preservatives) and the destroying microorganisms are important forms to prevent spoilage in food caused by bacteria.

4.3.2   PREVENTION OR DELAY OF FOOD BREAKDOWN

The destroying or inactivating their enzymes (scalding) and preventing or delaying chemical reactions (avoiding oxidation with the use of antioxidants) are some ways to prevent spoilage of food.

4.3.3   PREVENTING INJURIES CAUSED BY INSECTS, HIGHER ANIMALS, AND MECHANICAL CAUSES

To protect food against damage by microorganisms a maximum of the latency phase (a-b) and positive acceleration (b-c) of the growth curve must be prolonged (Figure 4.1). This is possible by ensuring that the smallest possible number of microorganisms reach the food the smaller the number of microorganisms, the greater the latency phase. Also, contamination by germs of no active growth (in logarithmic phase), present in containers, machinery, and utensils should be avoided.

Finally, unfavorable environmental conditions for food germs (humidity, temperature, WA, pH, or redox potential), or the presence of microbial inhibitors must be ensured in food (preservatives). The higher the number of unfavorable conditions, the longer it will take to start growth [1].

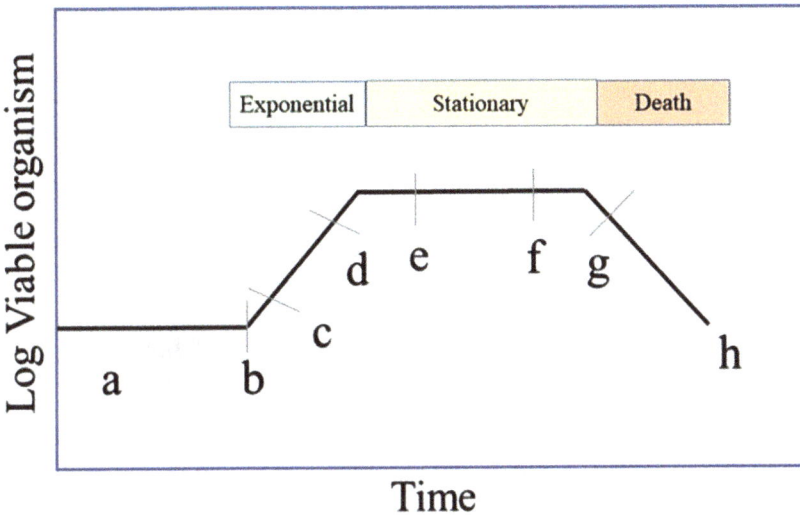

FIGURE 4.1   Growth curve of a bacterial population.
Note: where; a-b: latency; b-c: positive acceleration; c-d: logarithmic phase; d-e: negative acceleration; e-f: stationary phase; f-g: accelerated destruction; g-h: decline.

4.4   FOOD CONSERVATION METHODS

Preservation methods may be physical (low temperatures: refrigeration, freezing, high temperatures: pasteurization, sterilization, elimination of water: concentration, drying, freeze-drying, removal of air: vacuum packaging, packaging with CO_2 or N_2) or chemical like the use of food additives [4].

4.4.1   ASEPSIS

Asepsis is defined as freedom from pathogenic microorganisms' insufficient dose to cause an infection [5]. It is a way to prevent microorganisms from reaching the food. Applied mainly to raw foods. The type of microorganism (if it is pathogenic) and the number (size of the danger and the loading treatment) must be identified to know the potential risk.

4.4.2   ELIMINATION OF MICROORGANISMS

Microorganisms can be eliminated by filtration (impenetrable for bacteria), centrifugation, washing, and expulsion.

4.4.3   MAINTAIN ANAEROBIC CONDITIONS

Pathogenic microorganisms can be controlled by using vacuum closed containers, (for example, canned food). These microorganisms do not develop in the absence of oxygen, the same applies to temperature.

4.5   CONSERVATION USING HIGH TEMPERATURES

The destruction of microorganisms by heat is due to the coagulation of their proteins and especially to the inactivation of the enzymes necessary for their metabolism. Thermal sterilization technologies have been widely employed in the food industry since they are well developed and require low investment cost [6]. The heat treatment to choose to destroy vegetative cells and spores depend:

- Of other conservation methods to be used (to liven up or accentuate the applied temperature);
- The effects of heat on the food (should not destroy all its protein content, weaken the food).

4.5.1   HEAT RESISTANCE OF MICROORGANISMS AND THEIR SPORES

Heat resistance of microorganisms is usually expressed as a *time of thermal destruction*, which is the time necessary to destroy at a given temperature, a certain number of organisms under specific conditions. Figure 4.2 shows the resistance of microorganisms to heat treatment, is observed that some cells have low resistance (A-B), others have medium resistance (B-C), and other groups have high resistance (C-D), which are probably spores. Treatment must reach the entire population.

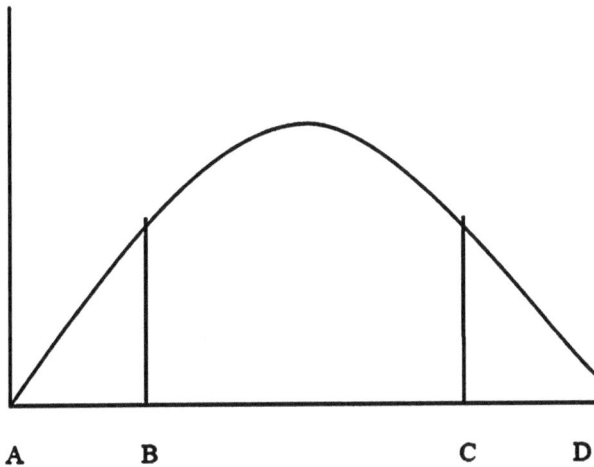

FIGURE 4.2   Cell resistance to heat treatment.

The factors that affect bacterial inactivation are:

1. **Temperature-Time Relationship:** The higher the temperature the less the destruction time.
2. **The Concentration of Spores (Cells):** The higher the number of spores, the more intense the heat treatment to destroy microorganisms.

3. **Preconditions for Bacteria and Spores:** The growth conditions of the bacteria and the production of spores, therefore as a subsequent treatment, will influence their resistance to heat:

   i. **Incubation Temperature:** Growth at optimal temperatures increases their resistance.

   ii. **Age:** Young bacteria are less resistant than mature bacteria.

   iii. **Desiccation:** Dissected spores are more difficult to destroy.

4. Composition of the substrate in which bacteria and spores are heated.

   i. **Humidity:** Moist heat is more potent to destroy (15–30 minutes at 121°C) than dry heat (3–4 hours at 160–180°C).

   ii. **pH:** They are more resistant to neutral pH; this resistance falls when the pH (acid) is lowered than when it rises (basic).

   iii. **Other Constituents of the Substrate:** NaCl in low concentrations have a protective effect on certain spores.

4.5.2 HEAT RESISTANCE OF YEASTS AND THEIR SPORES

Spores are the most resistant form of microbial cells and are well-equipped vehicles for colonization of food [7, 8]. The resistance to heat varies according to the different species and strains. In general, the spores are destroyed with barely 5–10°C or more of the temperature that would be needed to destroy vegetative forms. Ascospores are destroyed at 10–15 min/60°C [7, 8]; none resist brief heating at 100°C. Meanwhile vegetative forms are destroyed at 50–58°C/10–15 min.

4.5.3 THERMO RESISTANCE OF MOLDS AND THEIR SPORES

Spores are relatively stress-resistant structures but show a large variation in their ability to survive adverse conditions. Most spores are destroyed at 60°C/10–15 min. Asexual spores are more resistant than mycelia; 5–10°C or more is needed to destroy them. Fungal spores are quite resistant to dry heat; 120°C for 30 min is not enough to destroy them.

4.5.4   RESISTANCE TO BACTERIA AND THEIR SPORES

Some pathogens are easily destroyed, others need temperatures of 80–90°C for several minutes. Coccus is generally more resistant than bacilli. The higher the optimum temperature and maximum growth, the higher the heat resistance. Bacteria that have a capsule or form granules are more difficult to destroy, also because of their high lipid content.

4.5.5   THERMO ENZYME RESISTANCE

Most of the enzymes in food or microorganisms are destroyed at 79.4°C.

14.6   HEAT PENETRATION

The penetration of heat in food can be done by conduction (from molecule to molecule), convection (by the movement of liquids or gases), or radiation (energy emitted by molecules). The factors that determine the time required to raise the temperature of the center of the container to sterilization temperature are the following:

1. **Packaging Material:** Heating is slower in jars than in cans.
2. **Size and Shape of the Container:** The volume of the container should be considered.
3. **Initial Food Temperature:** Microorganisms must be longer in a lethal temperature from the beginning; however, it will not decrease the destruction time.
4. **The Temperature of the Autoclave:** This is necessary to reach the lethal temperature of the microorganisms.
5. **Consistency of can Contents, Shape, and Size of Food Portions:** Larger pieces are more quickly cooked than smaller ones.
6. **Rotation and Agitation:** Accelerate the heat penetration into the food and completely into the fluid.
7. Artificial rapid cooling is recommended because it is easily adjustable. Cooling too slowly can lead to overcooking of the food and allow the growth of thermophilic germs.

4.6.1   THERMAL DEATH RATE

To determine the heat treatment, one of these three methods is used:

- Graphical method;
- Mathematical formula; and
- Nomogram method.

The basic aspects of the mathematical formulas are presented below, for more details the book of Albert and Barbosa-Cánovas can be consulted. The death rate for any microorganism in a determined medium and thermally treated at certain fixed temperature follows first-order kinetics. Thus, if N is the number of microorganisms, its variation with time is expressed according to Eqn. (1):

$$\frac{dN}{dt} = -kN \tag{1}$$

This equation can be integrated, yielding:

$$N = N_0 exp(-kt) \tag{2}$$

where; N is the number of microorganisms present at a time t; and k is the death rate constant. The value of the rate constant depends on the type of microorganism, the medium, and the temperature [9].

Decimal reduction time (D) is the treatment time required to reduce the number of microorganisms to the tenth (10) part = D_T.

$$D_T = \frac{2.303}{k} log_{10}\left(\frac{N}{N_0}\right) \tag{3}$$

and since $N = 0.1\ N_0$, the decimal reduction time is expressed as a function of the rate constant of thermal death as:

$$D_T = \frac{2.303}{k} \tag{4}$$

and treatment time is expressed according to Eqn. (5):

$$t = D_T\ log_{10}\left(\frac{N}{N_0}\right) \tag{5}$$

4.7   THERMAL TREATMENTS EMPLOYED IN FOOD PROCESSING

The temperature and the time of the treatment of food will depend on the effect of the heat that exerts on the food and on other methods of conservation that will be used together. Some foods can be heated to a certain limit. The greater the heat treatment, the greater the number of germs destroyed until it reaches the temperature that causes the sterility of the product. The treatment must at least destroy those microorganisms with potential danger. In canned foods, all microorganisms that may alter the food during the last stages of manufacture must be destroyed. The different degrees of cooking used in food are classified as pasteurization, cooking around 100°C and cooking above 100°C.

4.7.1   PASTEURIZATION

Pasteurization is a heat treatment that destroys the microorganisms present in a food (generally liquid), this process is carried out at temperatures below 100°C. The cooking is employing steam, hot water, dry heat, or electric currents. Always cooled quickly after heating. Pasteurization is used when:

- Higher heat treatments would damage the qualities of the product (e.g., milk);
- One of the objectives is the destruction of pathogenic germs (e.g., milk);
- The most important alteration agents are not heat resistant (e.g., the yeasts of fruit juices);
- Surviving microorganisms are controlled with other additional methods (commercial milk cooling).

Other storage methods such as refrigeration (e.g., milk), packaging, and the addition of additives are used as a complement to pasteurization.

Pasteurization times and temperatures depend on the method used in the product to be treated:

- High temperature, short time: 71.1°C 15 sec;
- Short temperature, long time: 62.8°C/30 min.

4.7.2   HEATING AT APPROXIMATELY 100°C

Heating to temperatures close to 100° C is enough to destroy all microorganisms, except bacterial spores. This temperature is common in home-made preserves.

4.7.3   HEATING ABOVE 100°C

Temperatures above 100°C, are reached in steam autoclaves under pressure. Autoclave temperatures increase with increasing vapor pressures. The acronym UTH (ultra-high temperature) means that the food (usual milk) was heated to a temperature between 135 and 150°C (for 1–10 s) [10] by heat injection followed by the instantaneous evaporation, condensation, and fast cooling. Maintaining for enough time, this treatment capable of sterilizing the foods. These will have a long shelf life (6–9 months) without refrigeration [10, 11]. UHT treatment eliminates pathogenic bacteria and deactivates enzymes [10].

4.7.4   CONSERVATION USING LOW TEMPERATURES

The lower the temperature the slower the chemical reactions, enzymatic reactions, or microbial growth. A sufficiently low temperature will inhibit the growth of all microorganisms.

4.7.5   GROWTH OF MICROORGANISMS AT LOW TEMPERATURE

Freezing prevents the prolongation of most microorganisms. Refrigeration slows their growth rate. Temperatures of 5 or 6°C delay the multiplication of microorganisms producing food poisoning, except for *Clostridium botulinum* type E [12]. Microorganisms have been found growing up to –17°C.

4.7.6   TEMPERATURES USED IN STORAGE AT LOW TEMPERATURES

Low-temperature preservation is the most adopted method [13]. Low-temperature storage inhibits respiratory rate, spoilage reactions, and the growth of pathogenic microorganisms [3]. The most commercial storage units are below −18C.

4.7.7   REFRIGERATION

Consumers prefer refrigerated foods for freshness, low processing damage, and convenience when cooking [13]. Refrigeration is the main method of preserving food or as a temporary system until another form of conservation is applied. Cooling temperature, relative humidity, air velocity, and composition of the local atmosphere and possible use of UV radiation should be considered.

4.7.7.1   TEMPERATURE

The freezing temperature must be selected according to food class, time, and storage conditions. Some foods have optimal storage temperatures above freezing, for example, a banana should not be stored in the refrigerator, as it has a better storage temperature between 13.3–16.7°C; as for potatoes, they are better stored at 10–12.8°C.

4.7.7.2   RELATIVE HUMIDITY

Low humidity causes water loss and the wrinkling of food surfaces. If humidity is lower than the relative humidity of equilibrium, there is a water loss from the food to the exterior, leading to dehydration of the product [14]. High relative humidity favors the development of microorganisms causing alterations.

4.7.7.3   VENTILATION OR AIRSPEED CONTROL

Maintain a uniform relative humidity is important to eliminate odors and maintain food quality.

4.7.7.4   COMPOSITION OF THE STORAGE ATMOSPHERE

The amount of proportion of the gases in the atmosphere greatly influences. In the presence of optimal concentrations of CO_2 or O_3, the food remains unchanged for longer. Furthermore, higher relative humidity can be maintained, without jeopardizing the conservation and qualities of the food. This also favors the use of a storage temperature higher than the cooling temperature can be used.

4.8   IRRADIATION

The combination of UV irradiation with refrigeration helps to preserve several foods, allowing the use of humidity and higher temperatures than can be used in refrigeration. Some cheese and meat storage rooms are installing UV lamps.

4.9   FREEZING

Many microorganisms cannot grow in freezing [14]. Also, when food is frozen, part of the water is transformed into ice, thus decreasing the food's water activity (a_w). Many microorganisms cannot develop in low WA conditions [14].

4.9.1   ADVANTAGES THAT FAST FREEZING PRESENTS OVER SLOW FREEZING

Rapid freezing forms smaller ice crystals the mechanical so the destruction of food cells is scarce. Also, with a higher freezing speed, there is greater microbial and enzymatic inhibition. Fast freezing slows the chemical and enzymatic reactions of food by stopping microbial growth, the same effect produces intense or slow freezing.

4.10   DEHYDRATION

Drying is an excellent method of food preservation [15]. Fresh foods are very perishable because of their high moisture content. In food materials, water exists in two ways: both free water (intercellular spaces) and bound water (intracellular space). Free water is the solvent for microbial growth [15, 16]. Some examples are hot air drying, sun drying, vacuum drying, freezing drying [15, 16]. The factors that control dehydration are temperature, humidity, air velocity, and dehydration time. Dehydrated foods should be packed immediately after dehydration in suitable packaging. New research is being developed to develop new drying methods (more efficient and with less damage to food). Microwave drying [18], infrared (IR) drying, vacuum impregnation, ultrasound (US) assisted, lyophilization process [19], and osmotic dehydration are some of the new foods drying technologies [16].

4.11 NON-THERMAL (NON-CONVENTIONAL) PRESERVATION TECHNOLOGIES

The development of innovative non-thermal food processing technologies has received attention due to the increasing consumer demand for safe foods of high nutritional and functional quality [20]. In non-thermal preservation technologies, the temperature is not the main factor of the preservation of food (microorganisms and enzymes) [20, 21]. Some of the most promising emerging non-thermal technologies to extend food shelf life are pulsed electric fields (PEF), ultraviolet (UV) irradiation, high-pressure homogenization, high-pressure processing, IR heating, microwave heating, pulsed light (PL), ozone processing, and cold plasma (CP) [22]. Currently, non-thermal emerging technologies have been widely developed in rich countries. However, in America and Africa very few countries are investigating these technologies [20].

4.12 CONSERVATION BY ADDITIVES

4.12.1 FOOD ADDITIVES

A food additive is defined by Codex Alimentarius Commission's (which is recognized as the international standard) as "any compound not typically consumed as a food by itself and not normally used as an ingredient in the food but is intentionally added in the manufacture, processing, treatment, preparation, packing, transport, and holding of the food, to perform a technological function" [23]. The term does not include contaminants added to food for preserving or enhancing the nutritional potential [23].

4.12.2 ANTIMICROBIAL CONSERVATIVES

The antimicrobial conservatives increase the shelf life of the foodstuffs. However, the concentrations must be regulated so as not to affect the organoleptic properties of the food [24]. In general, antimicrobial conservatives must have a broad spectrum of antimicrobial activity, must not be toxic to humans and animals, should not affect the original taste of the food.

4.12.3 NATURAL FOOD ADDITIVES

Currently, there is a trend towards the use of natural additives, these are accepted by consumers for their perception as safer and healthier. Compounds derived from the secondary metabolism of most plants as polyphenols are recognized for having important biological properties such as antioxidant and antimicrobial activity [25]. Although consumers are leaning toward natural additives, this offer is still quite limited [26].

4.12.4 ADDITIONAL FOOD PRESERVATIVES

Additives not recognized in some regulations: Natural organic acids (lactic, malic, citric) and their derivatives; vinegar (acetic acid is a natural acid, more effective against yeasts and bacteria than against molds), NaCl, special sugars, their oils, CO_2, nitrous, and nitrates. These have been traditionally used for the preservation of meat products because of the effective antimicrobial action of nitrite against *Clostridium botulinum* [27]. The nitrates added to sausages and canned meats as antimicrobials can cause headaches [28]. However, the reduction of nitrates to nitrites (by nitrate-reductase in the bacterial flora) may cause acute toxic effects and the formation of carcinogenic substances due to the reactions between nitrogen oxide and amines [29].

Generally recognized as safe (GRAS): Propionic acid (affects the permeability of the membrane, used to prevent mold growth and formation of filamentous elements), caprylic acid, sorbic acid, and sorbates (inhibit yeasts and molds, lower against bacteria), K, Na, Ca, benzoic acid, benzoate, and other derivatives of benzoic acid (methyl and propyl esters of p-hydroxybenzoic acid). Essential oils [30].

KEYWORDS

- **asepsis**
- **generally recognized as safe**
- **metabolism**
- **microbial growth**
- **microorganisms**
- **postharvest**
- **ultra-high temperature**

REFERENCES

1. Tsironi, T., Houhoula, D., & Taoukis, P., (2020). Hurdle technology for fish preservation. *Aquaculture and Fisheries, 5*(2), 65–71.
2. Pernu, N., Keto-Timonen, R., Lindström, M., & Korkeala, H., (2020). High prevalence of *Clostridium botulinum* in vegetarian sausages. *Food Microbl., 91*, 103512.
3. Liu, D. K., Xu, C. C., Guo, C. X., & Zhang, X. X., (2020). Sub-zero temperature preservation of fruits and vegetables: A review. *J. Food Eng., 275*, 109881.
4. Sohail, M., Sun, D. W., & Zhu, Z., (2018). Recent developments in intelligent packaging for enhancing food quality and safety. *Crit. Rev. Food Sci. Nutr., 58*(15), 2650–2662.
5. Lucero, S., & Dryden, M., (2019). Antisepsis, asepsis and skin preparation. *Surgery, 37*(1), 45–50.
6. Li, X., & Farid, M., (2016). A review on recent development in non-conventional food sterilization technologies. *J. Food Eng., 182*, 33–45.
7. Dijksterhuis, J., (2019). Fungal spores: Highly variable and stress-resistant vehicles for distribution and spoilage. *Food Microbiol., 81*, 2–11.
8. Rozali, S. N. M., Milani, E. A., Deed, R. C., & Silva, F. V. M., (2017). Bacteria, mould and yeast spore inactivation studies by scanning electron microscope observations. *Int. J. Food Microbiol., 263*, 17–25.

9.   Barbosa, G., Fontana, J., Schmidt, S., & Labuza, T., (2007). *Water Activity in Foods* (1st edn.). Blackwell Publishing: State Avenue, Ames, Iowa, USA.

10. Zhang, D., Li, S., Palmer, J., Teh, K. H., Leow, S., & Flint, S., (2020). The relationship between numbers of *Pseudomonas* bacteria in milk used to manufacture UHT milk and the effect on product quality. *Int. Dairy J., 105*, 104687.

11. Pujol, L., Albert, I., Johnson, N. B., & Membré, J. M., (2013). Potential application of quantitative microbiological risk assessment techniques to an aseptic-UHT process in the food industry. *Int. J. Food Microbiol., 162*(3), 283–296.

12. Peck, M. W., & Van, V. A. H., (2016). Impact of *Clostridium botulinum* genomic diversity on food safety. *Curr. Opin. Food. Sci., 10*, 52–59.

13. Lei, J., (2012). Application of new physical storage technology in fruit and vegetable industry. *Afr. J. Biotechnol., 11*(25), 6718–6722.

14. Albert, I., & Barbosa-Cánovas, G. V., (2003). *Unit Operations in Food Engineering.* CRC Press.

15. Khan, M. I. H., Wellard, R. M., Nagy, S. A., Joardder, M. U. H., & Karim, M. A., (2017). Experimental investigation of bound and free water transport process during drying of hygroscopic food material. *Int. J. Therm. Sci., 117*, 266–273.

16. Qiu, L., Zhang, M., Tang, J., Adhikari, B., & Cao, P., (2019). Innovative technologies for producing and preserving intermediate moisture foods: A review. *Food Res. Inter., 116*, 90–102.

17. Torres, C., Rojas, R., Contreras, J., Serna, L., Belmares, R., & Aguilar, C., (2016). Mango seed: Functional and nutritional properties. *Trends Food Sci. Technol., 55*, 109–117.

18. Radoiu, M., (2020). Microwave drying process scale-up. *Chem. Eng. Process, 155*, 108088.

19. Sema-Cock, L., Torres-León, C., & Ayala-Aponte, A., (2015 Effect of adding non-caloric sweeteners on the physicochemical properties and on the drying kinetics of lyophilized mango peels. *Inf. Tecnol., 26*(4), 37–44.

20. Hernández-Hernández, H. M., Moreno-Vilet, L., & Villanueva-Rodríguez, S. J., (2019). Current status of emerging food processing technologies in Latin America: Novel non-thermal processing. *Innov. Food Sci. Emerg. Technol., 58*, 102233.

21. Roselló-Soto, E., Poojary, M. M., Barba, F. J., Koubaa, M., Lorenzo, J. M., Mañes, J., & Moltó, J. C., (2018). Thermal and non-thermal preservation techniques of tiger nuts' beverage "horchata de chufa". implications for food safety, nutritional and quality properties. *Food Res. Inter., 105*, 945–951.

22. Rifna, E. J., Singh, S. K., Chakraborty, S., & Dwivedi, M., (2019). Effect of thermal and non-thermal techniques for microbial safety in food powder: Recent advances. *Food Res. Inter., 126*, 108654.

23. FAO/WHO. (2021). *Food Additive Functional Classes.* http://www.fao.org/gsfaonline/reference/techfuncs.html (accessed on 28 September 2021).

24. Martins, F. C. O. L., Sentanin, M. A., & De Souza, D., (2019). Analytical methods in food additives determination: Compounds with functional applications. *Food Chem., 272*, 732–750.

25. Torres-león, C., Ventura-Sobrevilla, J., Serna-Cock, L., Ascacio-Valdés, J. A. J. A., Contreras-Esquivel, J., & Aguilar, C. N., (2017). Pentagalloylglucose (PGG):

A valuable phenolic compound with functional properties. *J. Funct. Foods, 37,* 176–189.

26. Carocho, M., Morales, P., & Ferreira, I. C. F. R., (2018). Antioxidants: Reviewing the chemistry, food applications, legislation and role as preservatives. *Trends Food Sci. Technol., 71,* 107–120.

27. Flores, M., & Toldrá, F., (2021). Chemistry, safety, and regulatory considerations in the use of nitrite and nitrate from natural origin in meat products. *Meat Science, 171,* 108272.

28. Gallo, M., Ferrara, L., Calogero, A., Montesano, D., & Naviglio, D., (2020). Relationships between food and diseases: What to know to ensure food safety. *Food Res. Inter., 137,* 109414.

29. Bedale, W., Sindelar, J. J., & Milkowski, A. L., (2016). dietary nitrate and nitrite: Benefits, risks, and evolving perceptions. *Meat Science, 120,* 85–92.

30. Falleh, H., Ben, J. M., Saada, M., & Ksouri, R., (2020). Essential oils: A promising eco-friendly food preservative. *Food Chem., 330,* 127268.

CHAPTER 5

Food and Diseases: What to Know in the Fight to Ensure Food Safety

CRISTIAN TORRES-LEÓN,[1] LEONARDO SEPÚLVEDA,[2] and
CRISTÓBAL NOÉ AGUILAR[2]

[1]*Reaserch Center and Ethnobiological Garden, Autonomous University
of Coahuila, Viesca – 27480, Coahuila, México,
E-mail: ctorresleon@uadec.edu.mx*

[2]*Bioprocesses and Bioproducts Research Group, Food Research
Department, Autonomous University of Coahuila, Saltillo, México*

ABSTRACT

This chapter talks about the importance of pathogenic microorganisms that affect food and that can cause disease in humans. General aspects of some pathogenic bacteria in food such as *Clostridium botulinum*, *Staphylococcus aureus*, *Salmonella*, *Clostridium perfringens*, *Escherichia coli*, and other indicators of fecal matter are mentioned. In addition, some non-bacterial pathogens are mentioned, such as some important viruses in the food area, fungi that can produce toxic secondary metabolites and parasites. Additionally, it talks about the importance of some microorganisms such as *Pseudomonas* related to the refrigeration and/or freezing process that some foods need.

5.1 INTRODUCTION

Food safety is a big problem for the human being, especially bacteria cause foodborne diseases. Since the beginning of mankind, foodborne diseases have been one of the leading causes of morbidity. More than 250 known and many unknown diseases are transmitted to humans mainly through the consumption of food contaminated with bacteria, fungi, parasites, or viruses.

5.2   HUMAN DISEASES TRANSMITTED BY FOOD AND WATER

Food and water may be the main vectors for the dissemination of infectious diseases [1]. Food contaminated by microorganisms or harmful substances causes infections, such as food poisoning and infections. Diseases caused by food or contaminated food constitute the most important vehicle of the pathogenic agent-oral pathway of penetration of the pathogen into the human organism. Foodborne illnesses can be divided into two broad categories: poisoning and infections (Figure 5.1).

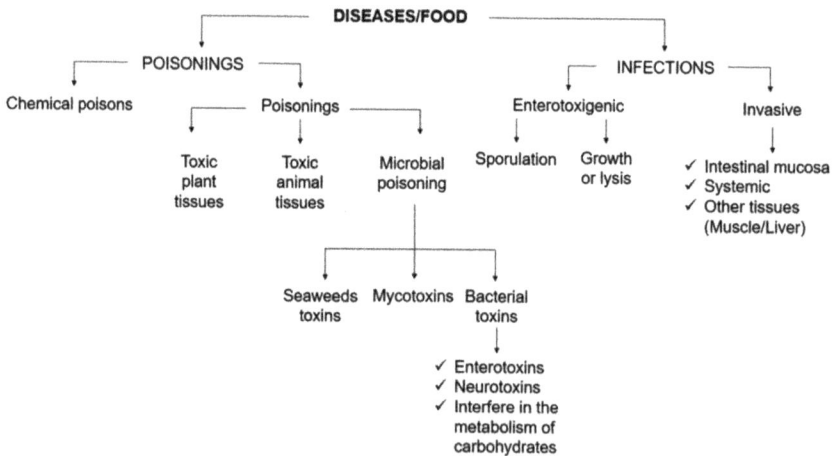

FIGURE 5.1   Poisoning and infections transmitted by food.

Different types of microorganisms can cause foodborne diseases, such as bacteria, viruses, parasites, protozoa, and fungi [1]. Food poisoning is caused by the consumption of foods containing the microorganisms or their toxins.

5.3 BACTERIAL FOOD POISONING

Around 66% of foodborne illness globally originates from pathogenic bacteria [2]. A disease caused by the presence of a bacterial toxin formed in the food. The disease symptoms are nausea, vomiting, abdominal pain, diarrhea, and fever [1]. Table 5.1 shows the main intoxications and infections caused by bacteria.

TABLE 5.1 Intoxications and Infections of Bacterial Origin

Intoxications	Infections
Staphylococcal: enterotoxin caused by *Staphylococcus aureus*	Salmonellosis: *Salmonella* endotoxin
Botulism: neurotoxin produced by *Clostridium botulinum*	Disease caused by *Clostridium perfringens*: enterotoxin released during sporulation
–	*Bacillus cereus* gastroenteritis: exoenterotoxin released during its lysis in the intestinal tract
–	Enteropathogenic *Escherichia coli* infection: several serotypes, some invasive and other enterotoxigenic
–	Other *Yersinia*, *Shigelose*, *Vibrio parahaemolyticus*

5.3.1 *BOTULISM*

One of the most serious diseases, caused by the ingestion of foods that contain a neurotoxin produced by *Clostridium botulinum*. Spores may find their way into foods via the environment [3]:

1. **Physiological Characteristics:** *C. botulinum* is an anaerobic, gram-positive, spore-forming bacillus that has an optimal temperature range of 30 to 37° C and a pH of 4.5 to 8.0. *C. botulinum* comprises four distinct groups of bacteria [4]. Groups I and II are associated with human botulism; Group III with botulism in animals, while Group IV has not been strongly associated with botulism [4]. There are seven confirmed *botulinum* neurotoxins (types A–G). The neurotoxins A, B, or E4 is most associated to foodborne botulism [4].

2. **Habitat:** The main habitats of *C. botulinum* are the animal intestinal tract, Food (fish, mice), and sediments of oceans and lakes.
3. **Dissemination:** The main form of dissemination is by asymptomatic carrier animals.
4. **Foods Involved:** *C. botulinum* spores exist widely in the environment and can contaminate food raw materials [5]. In general foods of low and medium acidity (not acidic), both home-canned and commercially processed vegetables are common sources of foodborne botulism, vacuum-packed food [5].
5. **Development and Production of Toxin:** It depends on the growth capacity of *C. botulinum* cells and food autolysis. Ideal water content to produce the toxin: 40% (30% inhibit production); 8% NaCl inhibits.
6. **Bacterial Growth:** pH below 4.5 prevents toxin formation.
7. **Toxin:** It is a protein with neurotoxin characteristics. It is absorbed by the small intestine and paralyzes the muscles. The inactivation of spores is achieved at 80°C/5–6 min, 90°C/15 min or 80°C/800 MPa [6]. Neurotoxins A, B, E, F, cause human diseases; neurotoxins C, D, diseases in animals (chickens, ducks, horses); lethal dose: 1 to 2 mg/kg.
8. **The Thermal Resistance of the Spores:** The spores have high resistance. General heat treatment applied 100°C/360 min, 105°C/120 min, 110°C/36 min, 115°C/12 min, 120°C/4 min (depending on the food).
9. **Symptoms:**
   i.   Before the development of neurological symptoms, gastrointestinal disturbances (nausea, vomiting, diarrhea, and stomachache);
   ii.   Visual disturbances;
   iii.   Paralysis of facial muscles, head, and pharynx;
   iv.   Respiratory paralysis;
   v.   High mortality rate.
10. **Incubation Period:** The incubation period is 12 to 48 hours.
11. **Duration of the Disease or Toxicological Process:** Lasts 3 to 6 days and recovery may take a few months.
12. **Control:**
   i.   Adequate heat treatment of food, ensuring total spore destruction;

ii. Use of chemical and physical methods (Aw reduction, acidification) that inhibit the development of the bacteria.

5.3.2 STAPHYLOCOCCAL POISONING

Disease caused by the ingestion of enterotoxin formed when certain strains of *Staphylococcus aureus* grow in food. The toxin was called enterotoxin because it causes gastroenteritis and inflammation of the gastric and intestinal mucous membranes. *S. aureus* is considered as a major cause of foodborne illnesses worldwide [7].

1. **Physiological Characteristics:** *S. aureus* is a typical representative of Gram-positive and facultative (anaerobes or aerobics) bacterium. This pathogen has catalase and coagulase positive [8]. The most important growth characteristics are:
    i.   **Optimum Temperature:** 35–37°C (10–45°C);
    ii. **pH:** 7.0 to 7.5 (4.2 to 9.3);
    iii. **High Salt Tolerance:** Tolerate media with 10 to 20% NaCl;
    iv. **Water Activity (Aw):** 0.99 to 0.86.
2. **Habitat:** In humans, the preferred colonization sites are the anterior part of the nostrils and the surfaces of the hands [9, 10]. In warm-blooded animals, *S. aureus* is found in the nasal mucosa, throat, and hair.
3. **Dissemination:** Food, animal, and biofilms serve as a virulence factor [11].
4. **Food Involved:** *S. aureus* can be found in various foods such as frozen birds, salads, raw meat, cakes, sweets with cream, and raw milk [11]. Recently, research has reported the presence of *S. aureus* in sushi and pork products [12].
5. **Infective Dose:** Pathogenic enterotoxins are usually produced when more than 10^5 CFU/g [8]; under these conditions, enough amounts of enterotoxins will be produced and released into the food causing poisoning when ingested.
6. **Enterotoxin:** *S. aureus* secretes multiple toxic proteins [8]. Designated, A, B, C1, C2, D, and E, based on their reactions with heat-resistant specific antibodies.
7. **Enterotoxin Dose:** 0.015 to 0.357 mg/kg body weight.

8. **Incubation Period:** The incubation period is 1 to 6 h.
9. **Symptoms:**
    i.   Nausea;
    ii.   Vomiting;
    iii. Diarrhea;
    iv.   Abdominal pain;
    v.   Some cases, intense salivation, sweating, and dehydration.
10. **Duration:** 24 hours, rarely fatal.

5.4 BACTERIAL FOOD INFECTION

5.4.1 *SALMONELLOSIS*

Salmonella is one of the main foodborne pathogens that affect humans and animals [13]. This pathogen can persist throughout the food supply chain thanks to their ability to form biofilms. Biofilms are communities of one or more species of bacterial cells locked in a self-produced polymeric matrix attached to a surface [14].

1. **Physiological characteristics:**
    - Gram-pods;
    - Facultative anaerobes;
    - Optimum temperature: 35 to 37°C (5 to 45°C);
    - Optimum pH: 6.5 to 7.5 (4.5 to 9.0);
    - Thermal resistance: low, 60°C/5 min;
    - Most known species: *S. typhi* and *S. paratyphi* A, B, C.
2. **Dissemination:** Food products as eggs are the principal source of *Salmonella* infections in humans. Other possible sources of *Salmonella* are bakery products, pork, milk products, beef, and various salads [14].
3. **Symptoms:** Food infection caused by *Salmonellae* is manifested by hyperacute gastroenteritis with abdominal pain, nausea, vomiting, mucous diarrhea (occasionally with blood), fever (38 to 39°C), and headache [1].
4. **Incubation Period:** The incubation period for *Salmonella* is 1 to 168 h (6 to 48 h).
5. **Duration:** The duration of the disease is 2 to 5 days.
6. **Infective Dose:** The effective dose of infection is 10^5 to 10^6 CFU/g.

5.4.2   CLOSTRIDIUM PERFRINGENS GASTROENTERITIS

C. perfringens is one of the most common pathogens [17]. This pathogen is a gram-positive, fermentative, non-motile, spore-forming, anaerobic rod-shaped bacillus [18]. The optimal growing conditions are temperature from 37 to 47°C and a pH from 5.5 to 8.0. The types *C. perfringens* are A, B, C, D, and E according to their ability to synthesize toxins.

1. **Habitat:** *C. perfringens* is commonly found in water, soil, and the intestinal tract of healthy animals [17, 18].
2. **Food Involved:** *C. perfringens* can contaminate, meat products (beef, chicken, and pork), seasonings, cakes, salty food, and fresh cheeses [15, 16].
3. **Infective Dose:** Large populations of *C. perfringens* cells (>6.0 log CFU, 10^5 vegetative cells per g) need to be ingested to cause foodborne illness [19, 20].
4. **Enterotoxin:** *C. perfringens* can produce up to 17 virulent toxins, which can cause infections [15]. Toxin can be inactivated at 60°C for 10 min.
5. **Symptoms:** The symptoms include diarrhea and severe abdominal pain, a large amount of gases, vomiting, and fevers, while nausea is less common [19].
6. **Incubation Period:** The incubation period is 8 to 24 h.
7. **Duration of the Disease:** The duration of the disease is 24 hours, usually of low severity.

5.4.3   ENTEROPATHOGENIC ESCHERICHIA COLI INFECTION

The bacterium *E. coli* inhabiting the intestine of warm-blooded animals. These bacteria coexist with other microorganisms assembling the commensal gut microbiota. However, some *E. coli* variants are pathogenic and can cause serious illness [21].

1. **Physiological Characteristics:** *E. coli* is a widely distributed gram-negative bacterium and is the most numerous facultative anaerobe [21]. *E. coli* ferment lactose with gas formation at 35°C. Optimal growing conditions are temperature 35 to 37°C and pH 6.5 to 7.5.

2. **Classification:**
   i.   *E. coli* classic enteropathogenic;
   ii. *E. coli* enterotoxigenic;
   iii. *E. coli* enteroinvasive;
   iv. *E. coli* enterohemorrhagic.
3. **Habitat:** *E. coli* habitant of the intestinal tract of humans and animals [22]. *E. coli* is one of the most common bacteria on farms.
4. **Dissemination:** Fecal material and through several vehicles can contaminate water and food. Toxigenic type, invasive type producing toxins, the infectious process of intestinal mucous membranes. Cattle are a major reservoir and source of *E. coli* and beef or beef products are commonly identified as a major source of *E. coli* disease in humans [23].
5. **Infection Dose:** The infection dose is 10^6 to 10^8 CFU/g.
6. **Incubation Period:** The incubation period is 6 to 36 h.
7. **Duration of Symptoms:** Mild watery or severe bloody diarrhea and complication such as hemolytic uremic syndrome, fevers, nausea, vomiting, and abdominal pains [23]. The duration of the disease is 1 to 3–4 days.
8. **Others:** Infection with *Vibrio parahemolyticus*, gastroenteritis due to *Bacillus cereus*, shigellosis due to *Shigella* spp., yersiniosis, due to *Yersinia* spp.

5.5 COLIFORM BACTERIA

Coliform bacteria are a heterogeneous group that habits the intestinal tract of both humans and animals [24]. There is little evidence to indicate that these coliform bacteria belong to a single taxonomic genus. The lack of certainty regarding its taxonomic affiliation and the imprecise correlation between the recommended methods for the detection of coliforms have generated problems. The first is that *Escherichia coli* is accepted as a coliform bacterium, the species contains variants that do not produce lactose gas or do so after 48 hours, so they are not identified by this technique. Second, the ability to ferment lactose is frequently associated with genes located in plasmids. These extrachromosomal characteristics are easily transferred between other Gram-negative bacteria unrelated to coliforms, which can, consequently, be recovered at the initial stage of the analysis. However, in practice, the technique has proven its effectiveness.

Enteric diseases caused by coliform bacteria are transmitted almost exclusively by fecal contamination of water and food. Transmission through contaminated water is one of the most serious sources of infection responsible for major epidemics of serious enteric diseases like typhoid fever and cholera. To evaluate the presence of pathogenic organisms in the water, the presence or absence of an organism and the population that is present in the water is determined, if pathogenic organisms are present, they indicate a contamination [25]. That organism is called an indicator.

This indicator must have the following characteristics:

- Apply to all types of water;
- Have a larger population in the environment than pathogens;
- Do not grow in water;
- Survive less than possible pathogenic organisms;
- Be detected through a simple and cheap methodology.

Unfortunately, there is no ideal indicator of sanitary water quality, but there are some organizations that approximate the requirements. To enable the comparison of data obtained in different locations and laboratories, the methodology must be standardized in the detection and the confirmation. All material used must be previously sterilized and sampling should be done in aseptic conditions. Transportation must be fast, and samples kept in ice baths. For chlorinated water samples, chlorine should be neutralized immediately after the sampling, with 100 mg/ml of sterile sodium thiosulfate solution.

5.5.1 INDICATORS OF FECAL POPULATION

- The most used fecal population indicator is the coliform group.
- They are typical fecal microflora organisms.
- *Escherichia coli* is of fecal origin only.
- The detection of coliform bacteria has been carried out in water since the end of the 19th century.
- The methodology of detection employs a selective medium (violet red bile agar) with Gram + inhibitors. The samples are mixed and incubated for 24 to 48 h. Then, colonies presumed to be positive are selected and inoculated into brilliant green lactose bile (24 h.). Gas evolution indicates that the food is coliform-positive [25].

- The incubation temperature is high, 35–37°C, for total coliforms, and 44.5 ± 0.2°C for fecal coliforms, they aim to prevent the growth of non-fecal bacteria more adapted to lower temperatures environment. For this reason, the analysis of fecal coliforms becomes more selective for *Escherichia coli*, and more specific to confirm contamination of fecal origin.
- The methodology may be used in bacteriological water examinations, for water measurement, for coliform group measurement, and should include two procedures: a multi-tube technique and a filter membrane technique.

The culture method needs 2 to 3 days to obtain the results. New research is being done to find methods of less time [25].

5.5.2   OTHER INDICATORS

No indicator is perfect and those intended to determine fecal contamination certainly do not function properly as population indicators of other origins. Coliforms and other fecal indicators can be supplemented with additional indicators that compensate for the inefficiency of the diversified population. Several groups of microorganisms have been shown suitable for that purpose: heterotrophic bacteria; yeasts; *Pseudomonas aeruginosa*; *Staphylococcus aureus*; viruses.

5.6   FOOD INTOXICATIONS AND FOOD INFECTIONS OF NON-BACTERIAL NATURE

5.6.1   VIRUS-TRANSMITTED DISEASES

Foodborne virus outbreaks carry both heavy public health. Food linked to outbreaks were leafy greens, soft berry fruits, fresh fruit, and shellfish [26, 27]. The symptoms include nausea, vomiting, and abdominal pain [26, 27]. Transfer of the coronavirus causing COVID-19 by food is low. Therefore, this virus is not considered a foodborne virus [27]. Gastroenteritis or hepatitis caused by human norovirus (NoV) and hepatitis A virus (HAV) is the main foodborne viral disease in humans. However, other viral agents have been implicated in food- and/or water-borne transmission of illness. Table 5.2 shows the main food-borne viral diseases.

TABLE 5.2 Virus-Transmitted Diseases

Disease	Agent	Food	Another Form of Transmission	Incubation and Symptoms	Prevention
Poliomyelitis	Poliovirus types I, II, III	Milk and other prepared drinks	Carrier disease, contaminated water	5–35 days; fever, vomiting, muscle pain	Personal hygiene, proper food storage
Infectious hepatitis	Infectious hepatitis virus	Milk, seafood, sweet potato salad	Carrier disease, contaminated water	10–50 days; loss of appetite	Seafood cooking
Acute non-bacterial gastroenteritis	Coxsackievirus	Possibly for food	Carrier disease, contaminated water	27–60 hours; fever, vomiting, and diarrhea	Personal hygiene, proper food storage

5.6.2 FUNGAL-TRANSMITTED DISEASES

5.6.2.1 MYCOTOXICOSIS

Mycotoxins are secondary metabolites produced by fungi that cause adverse, toxic, and pathological (carcinogenic) effects on humans and animals [28]. Mycotoxins can be hidden inside the fungal spores, in their mycelia and they will be released in food contaminated by these microorganisms. Mycotoxins occur mainly in cereals and oilseeds (rice, millet, wheat, barley, peanuts, cotton). US legislation applies a tolerance for aflatoxins in foods or cereals intended from 20 µg/Kg to 200 µg/Kg stipulated more by an economic concern than by public health. The maximum allowable concentration of aflatoxin (AFB1) in human food is 2 µg/kg [29]. Table 5.3 presents the main mycotoxins that affect food, this table also shows the main mycotoxin-producing fungi.

TABLE 5.3 Important Mycotoxins in Food

Mycotoxin	Fungi That Produce Them	Food	Affected Animals
Aflatoxin	*Aspergillus flavus, A. parasiticus; Penicillium*	Millet, cotton, oats, cocoa, rice, soy, wheat, cassava, potato	Cat, salmon, pheasant, rabbit, puppy, dog, rat, sheep, trout
Patulin	*Penicillium* and *Aspergillus*	Apple juice, watermelon	Rat, rabbit, shrimp
Ochratoxin A	*Penicillium* and *Aspergillus*	Millet, wheat, barley, chicken egg, cocoa	Rats, ducks
Luteoesquirina	*Penicillium islandicum*	Rice flour	Rat, trout
Sterigmatocystin	*Penicillium islandicum*	Wheat, oats	Rat
Penicillic acid	*Penicillium islandicum*	Tobacco	Rat
Toxic aleukia	*Cladosporium, Penicillium, Fusarium, Mucor, Alternaria*	Cereal grains	Human, rat
Roquefortine	*Penicillium roquefort*	Cheese	Rat

5.6.3 DISEASES TRANSMITTED BY PARASITES

Parasitic protozoa and helminths can cause disease with high morbidity in humans and animals [30]. In foods of non-animal origin, conventional pasteurization (72°C; 15 s) inactivates parasites (although some parasites are more heat resistant). Freezing at −20°C for 2 days inactivates most, but not all, parasites, and some are highly resistant to freezing [31]. Cooking at core temperature 60–75°C for 15–30 min inactivates parasites in most food of animal origin. Freezing at −21°C for 1–7 days generally inactivates parasites this parasite but cannot be relied upon in-home situations [32]. Table 5.4 shows the main diseases transmitted by parasites and the foods involved.

TABLE 5.4 Diseases Transmitted by Parasites

Causal Agent	Food	Incubation	Symptoms	Prevention
Anisakis spp.	Raw or under-cooked fish	Several days	Irritation of the pharynx and digestive tract	Cook the fish completely
Entamoeba hystolitica	Water contaminated with wastewater; wet food contaminated with human feces	From several days to 4 weeks	Diarrhea of different severity	Water supply protection
Taenia saginata	Raw or under-cooked beef	Several weeks	Abdominal pain, undefined discomfort	Cook beef completely
Diphyllobothrium	Raw or under-cooked fish	3 to 6 weeks	Generally, anemia in massive infestations	Cook fully or fish
Taenia	Raw or under-cooked pork	Several weeks	Digestive disorder, severe discomfort	Cook the pork completely
Trichinella	Pork or raw pork products, whale meat, seal	Nine days	Nausea, vomiting, diarrhea, fever, fatigue	Cook the pork completely; freeze the meat (−15°C/30 days)

5.7 PSYCHOTROPHS ASSOCIATED WITH THE DETERIORATION OF REFRIGERATED AND FROZEN FOODS

Psychrotrophs are cold-tolerant bacteria or archaea that can grow in temperatures of 0°C, but their optimum is around mesophiles: 20–30°C. The main Pathogenic psychrotrophs are *Clostridium botulinum* (type E, B, F), *Listeria monocytogenes*, *Vibrium cholerae*, *Yersinia enterocolitica*, Enteropathogenic *Escherichia coli*, *Aeromonas hydrophila*, *Pseudomonas aeruginosa* and *P. cocovenenans*.

5.7.1 PSEUDOMONAS

The genus *Pseudomonas* is capable of contaminating and spoiling protein-rich foods. This genus includes Gram-negative, non-fermentative, aerobic, catalase, and oxidase-positive, mesophilic, psychrotolerant, and non-spore-forming rods [33]. Generally, *Pseudomonas* produce water-soluble pigments, which can present fluorescence with ultraviolet (UV) light, for example, pyocyanin phenazine dye (Phenazines-deep green-bluish color). There are unpigmented lineages. Some species of *Pseudomonas* are important agents of diseases in plants and humans: *P. maltophilia, P. syringae, P. aeruginosa, P. cepacia, P. cocovenenans.* The most notable physiological property of these bacteria can use different organic compounds as carbon and energy sources. Among those compounds are alcohols, acids, amino acids, carbohydrates, and cyclic compounds. This extraordinary metabolic versatility demands a tremendous battery of special enzymes, inductive enzymes. Some species have optimal growth temperatures between 20 to 25°C (they are psychrophilic), and at 55°C, they are destroyed.

Pseudomonas generally metabolize hexoses (carbohydrates) through the Entner-Doudoroff pathway, which is well spread among bacteria, mainly Gram-negative bacteria, and rarely exists among anaerobes. This pathway is of great importance when the bacteria are growing in a medium rich in gluconate.

The aerobic degradation of the substrates generates CO_2, as the main or single product, through the Krebs cycle (tricarboxylic acid cycle). For the oxidation of innumerable substrates, the presence of innumerable special metabolic pathways, which converge for the Krebs cycle, is verified. Beta-ketoadipate pathway: Serves for the degradation of aromatic compounds and dicarboxylic acids, which are converted to acetyl CoA and succinate (intermediate of the Krebs cycle). Some of the substances used through this route are benzoate, p-hydroxybenzoate, toluene, phenol, naphthalene, catechol, and others.

5.7.2 IMPORTANT SPECIES IN FOOD SPOILAGE

Pseudomonas are found in a wide range of niches as soil, water, plants, animal tissues, and foods [34]. Pseudomonas species are predominant in fish due to their characteristics [33]:

- Psychrophiles, multiply at a cooling temperature;
- They can attack various substances in fish to produce odor-associated compounds (methyl mercaptans, dimethyl sulfide, dimethyl sulfate, dimethylamine, 3-methyl butanol);
- In the beginning, fresh fish has a flora, with a characteristic texture, with a predominance of *Flavobacteria*;
- During refrigeration, *Pseudomonas* grows easily. Approximately 9–10 days of storage, 60 to 90% of the population is *Pseudomonas*;
- In fresh fish *P. fragi* is in low percentage, but, with 12 to 15 days of storage, this species is responsible for 20–50% of the total contaminating flora.

Pseudomonas is one of the predominant genera in poultry, next to *Flavobacterium* and *Micrococcus*. After cooling, 90 to 95% of the microflora is from *Pseudomonas*. The range population of 10^8 cells/g has a bad smell.

Pseudomonas also affects foods such as milk since many enzymes (such as proteases) are resistant to heat and in UHT processes, proteases can cause coagulation and instability phenomena. Proteolysis of milk with refrigeration is initiated by *Pseudomonas* [33]. *P. fragi*, which produces a thermostable lipase that supports pasteurization when presented in raw milk. *Pseudomonas* and *Flavobacterium* cause silty and greening by fluorescent pigments and white dots. In meats at 10°C, *Pseudomonas* predominate. In dehydrated meats, *P. fluorescence* can also cause deterioration and release gas by denitrification. *P. syringae* causes the appearance of dark brown pigment in tomatoes, affecting its appearance and *P. fluorescens* and *P. ovalis*, produce fluorescent pigments in eggs. These are part of the flora causing the deterioration of the eggs.

KEYWORDS

- **aflatoxin**
- ***Clostridium botulinum***
- **foodborne**
- **hepatitis A virus**
- **microorganisms**
- **norovirus**

REFERENCES

1. Gallo, M., Ferrara, L., Calogero, A., Montesano, D., & Naviglio, D., (2020). Relationships between food and diseases: What to know to ensure food safety. *Food Res. Inter., 137*, 109414.
2. Toushik, S. H., Mizan, M. F. R., Hossain, M. I., & Ha, S., (2020). Do. fighting with old foes: The pledge of microbe-derived biological agents to defeat mono- and mixed-bacterial biofilms concerning food industries. *Trends Food Sci. Technol., 99*, 413–425.
3. Juneja, V. K., Purohit, A. S., Golden, M., Osoria, M., Glass, K. A., Mishra, A., Thippareddi, H., et al., (2020). A predictive growth model for *Clostridium botulinum* during cooling of cooked uncured ground beef. *Food Microbiol.*, 112490.
4. Peck, M. W., & Van, V. A. H., (2016). Impact of *Clostridium botulinum* genomic diversity on food safety. *Curr. Opin. Food Sci., 10*, 52–59.
5. Pernu, N., Keto-Timonen, R., Lindström, M., & Korkeala, H., (2020). High prevalence of *Clostridium botulinum* in vegetarian sausages. *Food Microbiol., 91*, 103512.
6. Lenz, C. A., & Vogel, R. F., (2015). Differential effects of sporulation temperature on the high pressure resistance of *Clostridium botulinum* type e spores and the interconnection with sporulation medium cation contents. *Food Microbiol., 46*, 434–442.
7. Rubab, M., Shahbaz, H. M., Olaimat, A. N., & Oh, D. H., (2018). Biosensors for rapid and sensitive detection of *Staphylococcus aureus* in food. *Biosens. Bioelectron., 105*, 49–57.
8. Zhao, Y., Xia, D., Ma, P., Gao, X., Kang, W., & Wei, J., (2020). Advances in the detection of virulence genes of *Staphylococcus aureus* originate from food. *Food Science and Human Wellness, 9*(1), 40–44.
9. Bencardino, D., & Vitali, L. A., (2019). *Staphylococcus aureus* carriage among food handlers in a pasta company: Pattern of virulence and resistance to linezolid. *Food Control, 96*, 351–356.
10. Parastan, R., Kargar, M., Solhjoo, K., & Kafilzadeh, F., (2020). *Staphylococcus aureus* biofilms: Structures, antibiotic resistance, inhibition, and vaccines. *Gene Rep., 20*. 100739.
11. Doulgeraki, A. I., Di Ciccio, P., Ianieri, A., & Nychas, G. J. E., (2017). Methicillin-resistant food-related *Staphylococcus aureus*: A review of current knowledge and biofilm formation for future studies and applications. *Res. Microbiol., 168*(1), 1–15.
12. Li, H., Tang, T., Stegger, M., Dalsgaard, A., Liu, T., & Leisner, J. J., (2020). Characterization of antimicrobial resistant *Staphylococcus aureus* from retail foods in Beijing, China. *Food Microbiol.*, 103603.
13. Lee, K. H., Lee, J. Y., Roy, P. K., Mizan, M. F. R., Hossain, M. I., Park, S. H., & Ha, S. D., (2020). Viability of *Salmonella typhimurium* Biofilms on major food-contact surfaces and eggshell treated during 35 days with and without water storage at room temperature. *Poult. Sci., 99*, 4558–4565.
14. Lamas, A., Regal, P., Vázquez, B., Miranda, J. M., Cepeda, A., & Franco, C. M., (2018). *Salmonella* and *Campylobacter* Biofilm formation: A comparative assessment from farm to fork. *J. Sci. Food Agric., 98*(11), 4014–4032.

15. Koo, B. S., Hwang, E. H., Kim, G., Park, J. Y., Oh, H., Lim, K. S., Kang, P., et al., (2020). Prevalence and characterization of *Clostridium perfringens* isolated from feces of captive cynomolgus monkeys (*Macaca fascicularis*). *Anaerobe, 64*, 102236.

16. Jang, Y. S., Kim, D. H., Bae, D., Kim, S. H., Kim, H., Moon, J. S., Song, K. Y., et al., (2020). Prevalence, toxin-typing, and antimicrobial susceptibility of *Clostridium perfringens* from retail meats in Seoul, Korea. *Anaerobe, 64*, 102235.

17. Hustá, M., Ducatelle, R., Haesebrouck, F., Van, I. F., & Goossens, E. A., (2020). Comparative study on the use of selective media for the enumeration of *Clostridium perfringens* in poultry faeces. *Anaerobe, 63*, 102205.

18. Wang, P., Huang, X., Yan, Z., Yang, Q., Sun, W., Gao, X., Luo, R., & Gun, S., (2019). Analyses of MiRNA in the ileum of diarrheic piglets caused by *Clostridium perfringens* type C. *Microb. Pathog., 136*(1), 103699.

19. Velugoti, P. R., Kumar, S., Bohra, L. K., Juneja, V. K., & Thippareddi, H., (2020). Inhibition of germination and outgrowth of *Clostridium perfringens* spores by buffered calcium, potassium and sodium citrates in cured and non-cured injected pork during cooling. *LWT, 123*, 109074.

20. Nakamura, H., Ogasawara, J., Monma, C., Hase, A., Suzuki, H., Kai, A., Haruki, K., & Nishikawa, Y., (2003). Usefulness of a combination of pulsed-field gel electrophoresis and enrichment culture in laboratory investigation of a foodborne outbreak due to *Clostridium perfringens*. *Diagn. Microbiol. Infect. Dis., 47*(3), 471–475.

21. Umpiérrez, A., Ernst, D., Fernández, M., Oliver, M., Casaux, M. L., Caffarena, R. D., Schild, C., et al., (2020). Virulence genes of *Escherichia coli* in diarrheic and healthy calves. *Rev. Argent. Microbiol.*

22. Sun, C., Wang, Y., Ma, S., Zhang, S., Liu, D., Wang, Y., & Wu, C., (2021). Surveillance of antimicrobial resistance in *Escherichia coli* and *Enterococci* from food products at retail in Beijing, China. *Food Control, 119*(2), 107483.

23. Grispoldi, L., Karama, M., Hadjicharalambous, C., De Stefani, F., Ventura, G., Ceccarelli, M., Revoltella, M., et al., (2020). Bovine lymph nodes as a source of *Escherichia coli* contamination of the meat. *Int. J. Food Microbiol., 331*, 108715.

24. Colclasure, V. J., Soderquist, T. J., Lynch, T., Schubert, N., McCormick, D. S., Urrutia, E., Knickerbocker, C., et al., (2015). Coliform bacteria, fabrics, and the environment. *Am. J. Infect. Control, 43*(2), 154–158.

25. Tominaga, T., (2019). Rapid detection of coliform bacteria using a lateral flow test strip assay. *J. Microbiol. Methods, 160*, 29–35.

26. Bosch, A., Gkogka, E., Le Guyader, F. S., Loisy-Hamon, F., Lee, A., Van, L. L., Marthi, B., et al., (2018). Foodborne viruses: Detection, risk assessment, and control options in food processing. *Int. J. Food Microbiol., 285*, 110–128.

27. Li, D., Zhao, M. Y., & Tan, T. H. M., (2020). What makes a foodborne virus: Comparing coronaviruses with human noroviruses. *Curr. Opin. Food Sci., 42*, 1–7.

28. Liao, Y., Peng, Z., Chen, L., Liu, L., Wu, Q., & Yang, W., (2019). Roles of microRNAs and prospective view of competing endogenous RNAs in mycotoxicosis. *Mutat. Res., 782*, 108285.

29. Zahran, E., Risha, E., Hamed, M., Ibrahim, T., & Palić, D., (2020). Dietary mycotoxicosis prevention with modified zeolite (Clinoptilolite) feed additive in Nile tilapia (*Oreochromis niloticus*). *Aquaculture, 515*, 734562.

30. Chalmers, R. M., Robertson, L. J., Dorny, P., Jordan, S., Kärssin, A., Katzer, F., La Carbona, S., et al., (2020). Parasite detection in food: Current status and future needs for validation. *Trends Food Sci. Technol., 99*, 337–350.

31. Gérard, C., Franssen, F., La Carbona, S., Monteiro, S., Cozma-Petruţ, A., Utaaker, K. S., Režek, J. A., et al., (2019). Inactivation of parasite transmission stages: Efficacy of treatments on foods of non-animal origin. *Trends Food Sci. Technol., 91*, 12–23.

32. Franssen, F., Gerard, C., Cozma-Petruţ, A., Vieira-Pinto, M., Jambrak, A. R., Rowan, N., Paulsen, P., et al., (2019). Inactivation of parasite transmission stages: Efficacy of treatments on food of animal origin. *Trends Food Sci. Technol., 83*, 114–128.

33. Caldera, L., Franzetti, L., Van, C. E., De Vos, P., Stragier, P., De Block, J., & Heyndrickx, M., (2016). Identification, enzymatic spoilage characterization and proteolytic activity quantification of *Pseudomonas* spp. isolated from different foods. *Food Microbiol., 54*, 142–153.

34. Franzetti, L., & Scarpellini, M., (2007). Characterization of *Pseudomonas* Spp. isolated from foods. *Ann. Microbiol., 57*, 34–47.

CHAPTER 6

Microbial Products of Importance in the Food Industry

JOSÉ L. MARTÍNEZ-HERNÁNDEZ,[1] JOSÉ SANDOVAL-CORTES,[2] and CRISTÓBAL NOÉ AGUILAR[1]

[1]Bioprocesses and Bioproducts Group, Food Research Department, Autonomous University of Coahuila, Saltillo, México, E-mail: jose-martinez@uadec.edu.mx (J. L. Martínez-Hernández)

[2]Analytical Chemistry Department, School of Chemistry, Autonomous University of Coahuila, Saltillo, México

ABSTRACT

Currently, the use of microorganisms to produce some secondary metabolite that has a high added value in the market and that additionally presents some biological activity has increased. This chapter talks about the use of enzymes produced by microorganisms and some biotechnological applications. The importance of microbial strains for fermentation processes applied in the food industry. In addition, the importance of the biomass of filamentous fungi, yeasts and bacteria is explained. General aspects are mentioned about the use of microorganisms to produce bioinsecticides. The importance of the production of antigens, antibiotics, ethanol, and other organic acids that are widely used in the food area are described. There is talk of the importance of some microorganisms to produce enzymes that are used in the food area. Finally, some general aspects are mentioned about the importance of bioconversion or biotransformation and the interest of the food industry to use these bioprocesses.

6.1   FOODS AND ENZYMES PRODUCED BY MICROORGANISMS

Microorganisms have been used in different processes and in different ways. Many substances of considerable economic value are products of microbial metabolism, from the industrial production of important materials including fine chemicals (pharmaceuticals) and those produced in large quantities that will be used as raw material [1].

6.2   BIOTECHNOLOGICAL APPLICATIONS OF MICROORGANISMS [1–3]

- Unicellular protein production;
- Insecticide production;
- Production of vaccines;
- Production of antibiotics;
- Ethanol production;
- Production of organic acids: Acetic acid (vinegar), propionic acid, citric acid and gluconic acid;
- Amino acid production: lysine, glutamic acid;
- Enzyme production;
- Solvent production;
- Polysaccharide production;
- Lipid production;
- Food production by lactic fermentation: acetoin, cheese, and yogurt.

6.3   PRODUCTION OF STRAINS FOR FOOD FERMENTATIONS

Microorganisms can be added for food fermentations as pure cultures or as a mixture of cultures. Or even if none is added, some foods to ferment, already contain the desired microorganisms in enough quantities (Table 6.1) [4].

In general, the principles of maintenance and preparation of crops are:

1.   **Selection:** They can be well characterized culture (from other laboratories) or selected after testing numerous species that must have stability, yield, and speeds at which they occur, should not present many variations.

2. **Maintenance of a Culture's Activities:** Once a satisfactory culture has been obtained, it must be kept pure and active, it can be through periodic transfers. Reserve cultures may be through lyophilization or freezing in liquid nitrogen.
3. Preparation of cultures.
4. **Activity of a Culture:** It is judged by the speed of its growth and production of substances.

TABLE 6.1 Microbial Production of Food Ingredients [4–8]

Ingredients	Function	Organism
D-arabitol	Sugar	*Candida diddensis*
Beta carotene	Pigment	*Blakeslea trispora*
Citric acid	Acidulant	*Aspergillus niger*
Diacetyl	Flavoring (butter)	*Leuconostoc cremosis*
Fatty acid esters	Fruit fragrances	*Pseudomonas* spp.
Decalactone	Fragrances	*Sporobolomyces odorus*
Geraniol	Rose fragrances	*Kluyveromyces lactis*
Glutamic acid	Flavor stimulant	*Corinebacterium glutamicum*
Lactic acid	Acidulant	*Estreptococos* and *Bacillus*
Lysine	Amino acid	*Corinebacterium glutamicum*
Mannitol	Sugar	*Torolopsis mannitofaciens*
Nisin	Antimicrobial	*Streptococcus lactis*
6-pentyl-alpha-pyrone	Coconut fragrances	*Trichoderma viridae*
L-phenylalanine	Precursor de aspartame	*Bacillus polymyxa*
Proline	Amino acids	*Serratia marcerscens*
Thermostable polysaccharides	Thickener	*Agrobacterium radiobacter*
Vitamin B12	Vitamin	*Propionibacterium*
Xanthan gum	Thickener	*Xanthomonas campestri*

6.4 MICROBIAL BIOMASS PRODUCTION/UNICELLULAR PROTEINS

Unicellular protein is that from microorganisms. Filamentous fungi, yeasts, bacteria, and algae are sources of unicellular proteins. Yeasts have been used more frequently, however this production depends on the type of substrate used, availability of the raw material, speed, and multiplication of the organism, toxicity, digestibility, and nutritional value of the final product [9].

6.4.1   FUNGAL BIOMASS [9–11]

- Filamentous fungi forming edible mushrooms:
    - For example, *Agaricus campestris*, *Lentinus* spp.; different lignocellulosic substrates are used to grow them (cane bagasse, rice straw, coffee pulp, cassava leaf);
    - Extract obtained from *Lentinus* spp., inhibition, or development of tumors in rats in research conducted at the University of Hong Kong;
    - Studies are currently being carried out on microbial protein for animal and human consumption, from agricultural waste used for fungal growth.
- Microprotein or fungal protein:
    - It is produced from the genus Fusarium, it is being studied as a meat substitute, it is considered a high-quality protein.
- Food yeast:
    - Yeast is a source of B-complex proteins and minerals. In the dry state, it contains about 50% good quality protein.

6.4.2   UNICELLULAR PROTEIN PRODUCTION PROCESS

- Abundant, low-cost, carbohydrate, industrial distillation waste, food processing, milk sera, lignocellulosic cane bagasse waste, old papers, are preferably used.
- It consists primarily of the transformation of carbohydrates from raw material into cellular matter, with supplementation of ammonium salts and phosphorus.
- Obtaining unicellular protein from yeast, the conversion of raw material into protein is conducted in fermenters that are equipped with agitation and aeration devices, trying to provide oxygen that is essential for the development of microorganisms.
- Once the conversion phase is finished, the recovery process is followed, which consists in separating the cell mass from the other

unwanted components from the culture medium, which is achieved by centrifugation, washing, filtering, and drying processes [9, 10].

6.5   PRODUCTION OF MICROBIAL INSECTICIDES-BIOLOGICAL CONTROL

The effects of pesticides on the environment have intensified efforts to use microorganisms as biological agents, rather than harmful chemicals for pest control. Microbial insecticides differ regarding the pathogenic characteristic that some microorganisms have for some insects. In each application the microorganisms need to be dispersed in the environment. Some specific requirements need to be filled before being used by microorganisms as microbial insecticides:

- Ease for large-scale production;
- The viability of the microorganism needs to be maintained;
- Organisms should not present toxicity or pathogenicity to animals and plants;
- The organism needs to be less expensive than chemical agents.

Bacteria, viruses, fungi, and protozoa can be used as insecticides. Some of them are produced commercially in large-scale fermentations or as insect hosts [10, 11].

There are about 400 species of fungi attacking insects and mites, which motivates studies based on the use of these fungi as fungicides and mycoherbicides.

The fungus *Metarhizium anisophae* is a natural enemy of the cicada leaf and can parasitize approximately 200 different host insects, and it has the possibility of being used as the sole biological agent controlling different pests.

More than 650 viruses have been characterized as plant pathogens and have demonstrated possibilities for their use as biological insecticides.

Three bacteria, *Bacillus popiliae*, *B. mortal*, *B. thuringensis* are commercially produced. *B. popiliae* is obtained directly from infected larvae, and the other bacteria are grown in submerged fermentation (SMF) processes. *B. thuringensis* is the most widely used bacteria and is grown in a medium that contains starch, corn syrup, casein, yeast extract and sucrose. After approximately 30 hours of fermentation, sporulation occurs accompanied by the formation of extracellular and intracellular proteins. *B. thuringensis* produces an intracellular protein crystal that has high toxicity, specific for larvae of certain lepidoptera insects, but is toxic to vertebrate animals and plants, after the crystal is ingested by the insect it is dissolved in the alkaline pH of the intestine, inhibiting ion transport, and thereby causing insect death [10–13].

6.6 PRODUCTION OF BACTERIAL VACCINES

6.6.1 IMMUNIZING ANTIGENS

An extensive culture of microorganisms for vaccine use can be viewed as an industrial process other than a fermentation process. Vaccines or antigenic substances of a protein nature previously prepared and presented by the microorganism itself or by products of its metabolism. All vaccines work by stimulating the organism to produce antibodies, substances that help the body defend against damage causing the development of immunity or a specific resistance. There are numerous damages caused by bacteria for which vaccines have been prepared: tuberculosis (*Mycobacterium tuberculosis* or Koch bacillus), meningococcal meningitis (*Neisseria menigitidis*), tetanus (*Clostridium tetanium*), among others [14, 15].

6.6.2 ANTIBIOTICS PRODUCTION

Antibiotics are a product of secondary metabolism that inhibits the growth process of other organisms, when used at low concentrations.

Penicillin was the first antibiotic to be produced industrially, currently approximately 6,000 known substances a minimum of 91 are commercially produced by fermentation. Another 46 semi-synthetic antibiotics have clinical applications. Chloramphenicol and pyrrole nitrin were originally discovered as microbial compounds and are now chemically produced in laboratories.

Microbial groups producing antibiotics:

- Antibiotics produced by fungi and bacteria.
- Fungi; only antibiotics produced by Ascomycetes (fungi) and imperfect fungi are of practical importance. Only 10 of the known

fungal antibiotics are commercially produced and only penicillin, Cephalosporin C, griseofulvin, and fusidic acid are clinically important.

- Bacteria; there are some important taxonomic groups that produce antibiotics. A wide variety of antibiotic structures and numbers has been found in actinomycetes especially in the Streptomyces genus.
- Another important group of substances are peptide antibiotics produced by the bacterium of the Bacillus genus.

The main phases of commercial production of penicillin (*P. chrysogenum*) are:

- Inoculum preparation;
- Preparation and sterilization of the medium;
- Inoculation of the non-fermenting medium;
- Sterile forced aeration during incubation;
- Removal of the mycelium after fermentation;
- Extraction and purification of penicillin.

The production of most antibiotics generally follows the same procedure. The main differences are related to the microorganism, the composition of the medium or the method of extraction. Some manufacturers use the same fermentation equipment to produce several different antibiotics [16, 17].

6.6.3 ETHANOL PRODUCTION

After water, alcohol is the most common solvent, in addition to representing the most widely used raw material in the laboratory and in the chemical industry. The biosynthesis of ethanol in selected strains of *Saccharomyces cerevisae*, perform alcoholic fermentation, from a fermentative carbohydrate. It is very important that a yeast culture has a vigorous growth and a high tolerance to ethanol, thus fermenting with a great final yield [18, 19].

Ethanol is inhibitor at high concentrations and yeast tolerance is a critical point for high production of this primary metabolite. Ethanol tolerance varies considerably according to the yeast's strain. Growth generally ceases when production reaches 5% ethanol (v/v), and the production rate is reduced to zero, a concentration of 6 to 10% ethanol (v/v).

The biochemical transformation performed by *S. cerevisae* is as follows:

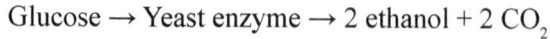

$$Glucose \rightarrow Yeast\ enzyme \rightarrow 2\ ethanol + 2\ CO_2$$

Ethanol can be produced from any fermented yeast carbohydrate: sucrose, fruit sugar, corn, molasses, beets, potatoes, malt, barley, oats, rye, rice, sorghum. It is necessary to hydrolyze complex carbohydrates into simple sugars, with the use of barley or fungal enzymes or by the heat treatment of acidified material. Cellulosic material, such as wood and paper pulp manufacturing waste can be used. Because of the large amount of cellulosic material residues available, direct fermentation of the materials when hydrolyzed by cellulolytic enzymes can be of great economic importance [18, 19].

Mixed cultures of *Clostridium thermocellum* and *C. thermosac-charolyticum* can be used. Hemicelluloses and celluloses are hydrolyzed into monosaccharides (hexoses and pentoses) by these bacteria and the monosaccharides are fermented directly to ethanol.

Ethanol production is initiated aerobically to produce the maximum biomass. In general, the process involves the following stages:

- Substrate preparation;
- Inoculum preparation;
- Fermentation;
- Distillation.

6.6.4   *ORGANIC ACID PRODUCTION*

6.6.4.1   ACETIC ACID PRODUCTION

In the food industry the group of acetic bacteria is of great importance in the manufacture of vinegars (alcohol, acetic acid) [21].
Group of acetic bacteria:

- Family Pseudomonadaceae, Gram negative, aerobic, mobile bacilli have a polar flagellum.
- They form acid by incomplete oxidation of sugars and alcohols (1), they form acetaldehyde by oxidation (2), acetaldehyde is

converted to acetic acid, 75% of acetaldehyde is converted into acetic acid and the remaining 25% in ethanol.
- They are reasonably tolerant of acidic conditions. They support a pH lower than 4 (optimal pH around 5 to 6).
- They cover the surface of plants, flowers, and fruits. Secondary flora of decomposing plant matter. They are quite nutritionally demanding.
- Main differences that distinguish them from the genus Pseudomonas:
  o Tolerate more acidic pH;
  o They have less proteolytic activity;
  o They have limited mobility and are not pigmented (except for *Gluconobacter oxydans* that produces a brown pigment).

Acetic bacteria can be divided into two main genera:

1. **Gluconobacter (Oxidize Glucose to Gluconic Acid):** *G. oxydans* presents polar flagella (3 to 8); they are considered sub-oxidative.
2. **Acetobacter:** *A. aceti, A. pasteurianus, A. peroxidans*: they are the most commercially used to produce vinegar.

They are undesirable contaminants in winemaking. Along with yeasts and lactic bacteria, acetic bacteria have been frequently cited as harmful in the beverage industry. The type of damage they produce includes unpleasant flavors, silty growth, and gas formation.

6.6.4.2   GENERALITIES ABOUT VINEGAR PRODUCTION

- Vinegar production involves two types of biochemical alterations:
  o Alcoholic fermentation of a carbohydrate; and
  o Oxidation of alcohol to acetic acid.
- There are various types of vinegars produced depending on the type of material used in alcoholic fermentation (fruit sugars, hydrolyzed starchy syrups).
- Yeast fermentation is needed for alcohol production. The alcoholic concentration is adjusted between 10 to 13% and the bacteria are then exposed to acetic acid (an aerobic process) that will oxidize the alcoholic solution to the vinegar in the desired concentration.

With the increase in the production of beverages bottled in plastic, non-fermentative acetic bacteria have become more important. Several reasons contribute to this among them the resistance of Gluconobacter to sanitizers commonly used in the beverage bottling industry, its ability to grow in the presence of ascorbic and benzoic acid, and the high levels that characterize beverages in plastic containers.

6.6.4.3 PRODUCTION OF CITRIC ACID

The accumulation of citric acid by some fungi was discovered in 1893 when Wehmer discovered that *Citromyces* (known today as *Penicillum* spp.) and Mucor possessed the ability to accumulate this acid during its cultivation [22].

Currently, industrial fermentation for citrate production is carried out using a single species of fungus: *Aspergillus niger*. The oxidative yeast *Saccharomycopsis* (*Candida*) has presented interesting characteristics being considered potentially important.

6.6.4.3.1 Citrate Use

- About 70% of the production is used by the food and beverage industry, 12% by the pharmaceutical industry and 18% by other industries.
- In the food industry, it is used on a large scale as an acidulant to present a pleasant taste, very low toxicity, and high solubility. This acid has the capacity of work with heavy metals such as iron and copper. This property has led to a growing use as an oil and fat stabilizer to reduce its oxidation catalyzed by these metals. Also, that property together with its low degree of corrosivity to certain metals, has allowed its use in the cleaning of boilers and special installations.
- In the pharmaceutical industry, citric acid is used as a stabilizer for ascorbic acid due to its chelating action. In effervescent antacids and analgesics, citric acid is used together with carbonates and bicarbonates to generate CO_2.
- Citrate salts, such as trisodium citrate and tripotassium citrate are used in medicine to prevent blood clotting and in the food industry as an emulsifier to produce cheese.

- Citric acid esters such as triethyl, tributyl, and acetyl dibutyl are used as non-toxic plasticizers in plastic food packaging films.

Biochemistry of the production of citric acid: Essentially glucose is transformed into pyruvate by the glycolytic pathway. Pyruvate is transformed into acetyl CoA, which will enter the Krebs cycle for citrate formation.

Source of carbohydrate to produce citric acid: The presence of an easily metabolizable carbohydrate is essential for a production of citric acid. Maltose, sucrose, mannose, glucose, and fructose are the most appropriate sugars for acid production. In practice, citric acid is produced from purified carbohydrate (sucrose) or from a source of crude carbohydrate, like sugar cane molasses, beet molasses, crude sucrose, cane broth, starch hydrolysate.

The presence of metals as contaminants of raw materials is the main problem in citrus fermentation. Some products contain inhibitors and/or growth promoters, most of them little known or analyzed. Some techniques are used to remove or neutralize inhibition by these contaminants: the addition of potassium ferrocyanide and methanol is quite common in practice.

6.6.4.4 LACTIC ACID PRODUCTION

To produce lactic acid, homolactic bacteria of the genus *Lactobacillus* and *Streptococcus* are used. The species chosen depends on the available carbohydrate and the temperature used [23, 24]:

- *L. delbrueckii*, *L. bulgaricus*: temperature between 40 to 50°C;
- *L. casei* and *S. lactis*: temperature around 30°C;
- *L. pentosis*, *L. leshmanii*: temperature above 30°C.

6.6.4.4.1 Main Characteristics of Lactic Bacteria

They are Gram (+) bacteria, microaerophilic, non-sporulated, usually do not have motility, are catalase negative, have small and unpigmented colonies. Nutritionally they are very demanding. They have limited biosynthetic ability and need amino acids, vitamins, purines, and pyrimidines. They need for their growth in the laboratory: peptones and protein hydrolysates, yeast extracts and nucleotides. They are acidophilic bacteria: they tolerate low pH values. Tolerate a pH: Bacilli: do not grow at a pH greater than 6.0; optimum pH for growth is 4.5; coccus: neutral pH. When they grow in the presence of O_2, oxidizing toxic substances of the metabolism are produced: peroxides (H_2O_2), superoxide (O^{-2}), and hydroxyl radical (OH-).

Habitat of Bacilli and homofermentative coccus:

- The main one is the body of warm-blooded animals.
- **Animal:** Normal flora of skin and mucous membranes: oropharynx, gastrointestinal tract, and genitourinary tract (Associated with pathogens, exclusively strains of the genus *Streptococcus*).
- **Vegetable:** They live in association with vegetables and grow remnants of nutrients removed from the vegetable.
- **Milk:** Access through animal skin and mucous membranes (fodder, vegetables), pathogenic strains (*Streptococcus agalactia*).

Lactic bacteria can be divided into two biochemical subgroups according to the products formed from glucose:

1. **Homofermentative Bacteria:** They are very important in the production of lactic acid. The first steps of the lactic fermentation metabolic pathway are the same as those of alcoholic fermentation, or more specifically of the Embden-Meyerhof pathway or glycolytic pathway. An important intermediate for the formation of lactic acid is pyruvic acid. At the end of the glycolytic pathway, pyruvic acid, under the action of the enzyme lactate dehydrogenase gives rise to lactic acid.

2. **Heterofermentative Bacteria:** The fermentation of glucose by these bacteria results in several products. As soon as homofermentative bacteria degrade glucose through the glycolytic pathway, heterofermentatives degrade glucose through the oxidative pathway of pentoses phosphate. The most important intermediary components of the heterofermentative pathway are pyruvic acid and acetaldehyde. The liquid yield in ATP: 2 moles/mol of glucose by the homofermentative route and just 1 mol/mol of glucose by the heterofermentative route.

6.6.4.4.2 Process for Obtaining Lactic Acid

Lactic acid is obtained from various raw materials, waste by-products from the food industry, such as cheese whey, molasses, corn glucose. Waste of high BOD (biochemical oxygen demand) is also used, such as those of the pulp and paper industry, agglomerates containing sugar polymers. The substrates used are mainly glucose, lactose, and sucrose. For example, corn, potato, and cassava starchy substrates can be used, since they are pre-hydrolyzed enzymatically. The concentration of sugars in the sample is adjusted between 5–20% according to the microorganism, the raw material and the process used:

1. **pH:** To promote a high yield should be approximately neutral or slightly acidic. It is important to maintain a constant pH, because as acidity increases an inhibition of fermentation occurs.
2. **Fermentation Time:** Fermentation is completed between 1 to 7 days (5–7 days).
3. **Yield:** Approximately 85–90% in relation to the sugar consumed (fermented). The acid formed is a racemic mixture.

6.6.4.4.3 Use of Lactic Acid

- It is used in food, fermentations, pharmaceuticals, cosmetics, and in the chemical industry.
- In food, it is used as an acidulant in confectionery products, in the manufacture of extracts, essences, fruit sugar, refrigerants, and others. It is also used as a curator of meat, vegetables, and fish.
- In the textile industry it is used as a mordant for printing. It is also used in the preparation of leather. In the manufacture of plastic, transparent lactic acid of superior quality is used.
- Lactates are used in the pharmaceutical, cosmetic, and food industries.
- Esters are used mainly in the manufacture of dyes and varnishes, plasticizers, and as solvents.

6.6.4.4.4 Lactic Fermentation

Bacteria used in the industry are anaerobes and microaerophiles, to produce acetic, lactic, gluconic, propionic, and other acids and to produce foods such as cheeses, vinegars, milks, fermented, and others.

Fungi are also used in the production of acids by fermentation. The main acids are citric, gluconic, fumaric, lactic, gallic, fatty acids and others.

The bacteria involved in the processes for obtaining acids are mainly from the genus Acetobacter and Lactobacillus. Bacteria can form innumerable different acids. Therefore, some of the bacteria that produce lactic acid, acetic acid and propionic acid are of a great economic interest. Acids are derived from anaerobic degradation of glycerol by incomplete oxidation.

6.6.4.4.5   *Importance of Lactic Bacteria in the Food Industry*

1. **Obtaining Fermented Vegetables:** Acetones, fodder for cattle.
2. **Genus Leuconostoc:** Production of flavor in dairy products: yogurt, acidified milks, cheeses, and butter. Leuconostoc, *S. lactics*, *S. diacetilactis* and *L. cremoris*: they are used as sources of flavorings in the dairy industry and are responsible for the different characteristics of butter, cheese, and yogurt (diacetyl production).
3. **Cured Meats:** Salamis and other sausages.
4. **Biopolymers:** Thickeners; plasma expander (plasma to replenish the volume in large hemorrhages).
5. **Flavors:** Production of diacetyl/ketone from citrate in milk 1 g/L.

6.6.4.4.6   *Negative Aspects of the Presence of Lactic Acid Bacteria (LAB) in the Industry*

- Production of acidity and undesirable aromas (diacetyl) in wines, sugars, beers, and other distilled beverages, for example, *Pediococcus perniciosus* and *P. damnosus* found in beer.
- Deterioration of meat products, vegetables, and fruits.
- The synthesis of biopolymers by *Leuconostoc mesenteroides*, consumes sucrose in the sugar industry reducing the yield and causing the blocking of filters, pumps, and pipes.

6.6.4.5   *PRODUCTION OF PROPIONIC ACID [25]*

1. **Microorganism:** Propionibacterium.
2. **Raw Materials:** Carbon source: lactose, sucrose, and glucose; Nitrogen source: myosin and peptone. Thiamine and riboflavin stimulate the growth of propionic bacteria.

3. Process:

Glucose → glycolysis → pyruvic acid → succinic acid decarboxyl-
ation → Propionic acid

6.6.4.6 *PRODUCTION OF GLUCONIC ACID*

Many microorganisms can synthesize amino acids from inorganic nitrogen
compounds and the amount of synthesis of some amino acids can exceed
the cellular needs for protein synthesis resulting in their excretion into the
environment [26].

Some microorganisms can produce enough amounts of amino acids
(lysine, glutamic acid, and tryptophan) to justify their commercial use.
Many species of microorganisms, especially bacteria and fungi, can
produce large amounts of glutamic acid.

The species of the genera *Micrococcus*, *Arthrobacter*, and *Brevibacte-
rium* are used in the industrial production of glutamic acid (they produce a
minimum of 30 g of amino acid/L of medium). The production and excretion of glutamic acid depends on cell permeability and the increase in cell
permeability can be obtained by:

- Biotin deficiency;
- Addition of saturated fatty acids or fatty acid derivatives;
- Addition of oleic acid;
- Glycerol deficiency for glycerol;
- Adding penicillin in the logarithmic phase of growth.

All strains producing glutamic acid need biotin, an essential coenzyme
in the synthesis of fatty acids. Biotin more than 5 mg/L, increases the
synthesis of oleic acid and results in a membrane with high phospholipid
content, causing cells unable to excrete the glutamic acid formed (25–35
mg of glutamic acid/mg of dry mass) and accumulated intracellularly.

The culture medium to produce glutamic acid generally contains a
carbohydrate, peptone, inorganic salts, and biotin in a suitable concentration. Alpha-ketoglutaric acid, intermediate of the Krebs cycle and/or
precursor of glutamic acid. The conversion of alpha-ketoglutaric acid to
glutamic acid is carried out by the action of glutamic dehydrogenase.

One of the main uses of glutamic acid is as a flavoring agent and
flavoring agent, in the form of sodium glutamate (Table 6.2).

TABLE 6.2   Enzyme Production from Microorganism [27]

Enzymes	Origen	Industry	Application
Amylase	*Asperdillus niger, A. oryzae, Bacillus subtilis, Rhizophus spp., Mucor rouxii*	Bakery	Flour supplement, dough preparation, pre-cooked food, syrup making
Cellulose	*Aspergillus niger, Trichoderma viridae*	Beer	Preparation of coffee concentrates, sugar clarification
Dextran sucrose	*Leuconostoc mens.*	Food	Dextran for various uses
Glucose oxidase	*Aspergillus niger*	Food	Removal of glucose from egg solids
Invertase	*Saccharomyces cerevisiae*	Food	Artificial honey
Lactose	*Saccharomyces fragilis*	Dairy	Lactose hydrolysis
Lipase	*Aspergillus niger, Rhizopus spp., Mucor spp.*	Dairy	Cheese flavor
Pectinase	*Aspergillus niger, Rhizophus spp., Penicillium*	Food	Wine and fruit sugar clarification
Protease	*Aspergillus oryzae, Bacillus subtilis*	Beer, bakery, food	Prevents beer from clouding on cooling, meat tenderizer
Enzymes like renin	*Mucor*	Food	Cheese curd for cheese making

6.7   BIOCONVERSIONS OR BIOTRANSFORMATION

They are processes in which microorganisms convert a compound to structurally related products. They comprise only one or few enzymatic reactions (unlike fermentative processes that have several reaction sequences) [28]:

- They are used commercially when conventional chemical reactions are very expensive or difficult. For example, stereoselective transformations, with just one group of a molecule, with several identical functional groups need to be modified.
- Bioconversion, involves the growth of the organism in large fermenters, followed by the addition at an appropriate time of a chemical compound to be converted.

- The bioconversion process is mostly applied in the production of hormones, steroids.

6.7.1   STEROID BIOCONVERSION-CORTISONE SYNTHESIS

- Corticosteroids (cortisone) and ACTH are used with good effect in the treatment of rheumatoid arthritis.
- Steroids are important hormones in animals, they regulate various metabolic processes. They can be obtained from plants and animals.
- Many microorganisms can transform steroids; however, fungi are the most important.
- Some steroids are used in human medicine: Progesterone is activated during pregnancy; some steroids are used as sedatives in antitumor therapy, cortisone is anti-inflammatory, rheumatoid arthritis, skin damage and allergies.
- The transformation of steroids requires the growth of a micro-organism capable of effecting the desired modification, in an appropriate medium and under controlled conditions. After growth has occurred, the steroid is added to the culture and the chemical transformation is carried out during a subsequent incubation. A mass of microbial development is removed, then the transformed steroid is removed.

6.7.2   FUNGAL STEROID TRANSFORMATION [29]

- Reduction of double ligatures in positions 4,5 and 16,17 and reduction of ketone groups, for example, *Epicoccum oryzae.*
- Steroid dehydrogenation at position 1,2, for example *Fusarium solani.*
- **Hydroxylation:** It is the most important transformation. In positions 11 and 17 it is quite productive. Hydroxylation in more than one position. Progesterone in positions 6 and 11 by *Rhizopus arrhizus.*
- **Epoxidation:** *Curvularia lunata.* Growth in a minimum medium, reduces problems with product extraction. At the end of growth, the steroid is added and dissolved in small amounts of solvent

(0.5–2 g/L). After 20–40 hours, the extraction is performed with chloroform solvent.

Examples of bioconversions carried out by microorganisms [28, 29]:

- Synthesis of vitamin C (*Acetobacter suboxydans, A. xylinum*);
- Transformation of semi-synthetically produced antibiotics, for example, enzyme penicillin acylase (fungal) used in the production of penicillin. Isomerization of glucose to produce fructose (enzyme glucose isomerase);
- Transformation of soil-polluting pesticides and effluents, related genera: *Arthrobacter, Phanerochaete chrysoporium, Pseudomonas cepacia, P. aeruginosa*, among others.

6.8   INDUSTRIAL INTEREST

Fungi influence the life of humans by participating in desirable and harmful processes. Fungi are widely found in all ecosystems and habitats. They can be parasites, symbionts, being mostly saprophytes. They grow where there is organic matter available, whether alive or dead, usually where there is heat and humidity. Water, soil, logs, leaves, fruits, seeds, excrement, insects, fresh, and processed foods, and innumerable human-made products constitute substrates for the development of men.

6.8.1   FUNGI IN BIOTECHNOLOGY [4–6, 31]

Many species of fungi have been studied and used to produce substances of industrial or medical interest. Ethanol, citric acid, gluconic acid, amino acids, vitamins, nucleotides, and polysaccharides are examples of primary metabolites produced by fungi in that antibiotics are important secondary metabolites.

Beyond the application of fermentation, new biotechnological aspects have been explored, including environmental ones, that is, fungi can act as beneficial agents for the improvement of the environment: treatment of liquid waste and bioremediation of contaminated soils, mineralogy, and biohydrometallurgy, biomass production, including edible proteins, fuel technologies, particularly coal solubilization; Biological control employment.

6.8.2   YEASTS

Yeasts are fungi like filaments, but they differ from them by being single-celled. Because they are simpler cells they grow and reproduce faster than filamentous fungi. A typical yeast consists of oval cells that commonly asexually multiply by budding. Most yeasts do not live in the soil are better suited to environments with high concentrations of sugars, such as flower nectar and fruit surface.

Fermentative yeasts have been explored by man for thousands of years, in the production of beer and wine and in the fermentation of bread, however, only in the 19th century has the biological nature of the agents responsible for these processes The main agent of alcoholic fermentation, *Saccharomyces cerevisiae*, is an ascomycete yeast [31].

6.8.2.1   YEASTS OF INDUSTRIAL INTEREST

Saccharomyces cerevisiae, S. calrsbergensis are used in baking, beer, wines; *S. fragilis, S. lactis* ferment lactose (waste treatment); *S. roufii, S. mellis* are osmophilic, dried fruits, syrups, jellies; *S. baillie*, in the fermentation of sugars (citrus).

Osmotic *Torulopsis*-condensed milk, Candida produces large amounts of protein, attacks milk and its derivatives; *Rodutorulla*-belongs to deteriorating meats (red or yellow) *Picchia, Hansenula, Debarymocyces, Thrichosporum*, in deterioration and as a film production, oxidizes acetic acid and alters the taste, *Debaryomyces*-meat, cheese, and sausages [31].

KEYWORDS

- **biochemical oxygen demand**
- *Clostridium tetanium*
- *Lentinus* **spp.**
- *Metarhizium anisophae*
- **microorganisms**
- **microprotein**

REFERENCES

1. Adrio, J. L., & Demain, A. L., (2014). Microbial enzymes: Tools for biotechnological processes. *Biomolecules, 4,* 117–139.
2. Kirk, O., Borchert, T. V., & Fuglsang, C. C., (2002). Industrial enzyme applications. *Curr. Opin. Biotechnol., 13,* 345–351.
3. Talon, R., & Zagorec, M., (2017). Special issue: Beneficial microorganisms for food manufacturing-fermented and biopreserved foods and beverages. *Microorganisms, 5,* 71.
4. Tamang, J. P., Watanabe, K., & Holzapfel, W. H., (2016). Review: Diversity of microorganisms in global fermented foods and beverages. *Front Microbiol., 7,* 377.
5. Bourdichon, F., Casaregola, S., Farrokh, C., Frisvad, J. C., Gerds, M. L., Hammes, W. P., Harnett, J., et al., (2012). Food fermentations: Microorganisms with technological beneficial use. *Int. J. of Food Microbiol., 154,* 87–97.
6. Pometto, A. L., & Demirci, A., (2006). Technologies used for microbial production of food ingredients. In: *Pometto, A., Shetty, K., Paliyath, G., & Levin, R. E., (eds.), Functional Foods and Biotechnology* (pp. 1–12*). CRC Press.*
7. Longo, M. A., & Sanromán, M. A., (2006). Production of food aroma compounds: Microbial and enzymatic methodologies. *Food Technol. Biotechnol., 44,* 335–353.
8. Bhargav, S., Panda, B. P., Ali, M., & Javed, S., (2008). Solid-state fermentation: An overview. *Chem Biochem. Eng., 2,* 49–70.
9. Sanaa, O., & Soraya, S., (1991). Microbial biomass and protein production from whey. *Med. J. Islamic World Acad. Sci., 4,* 170–172.
10. Luna-García, H. A., Martínez-Hernández, J. L., Ilyina, A., Segura-Ceniceros, E. P., Aguilar, C. N., Ventura-Sobrevilla, J. M., Flores-Gallegos, A. C., & Chávez-González, M. L., (2020). Production of unicellular biomass as a food ingredient from agro-industrial wasted. In: Zakaria, Z. A., Boopathy, R., & Dib, J. R., (eds.), *Valorization of Agro-Industrial Residues: Biological Approaches* (Vol. I, pp. 219–238). Springer.
11. Sibtain, A., Ghulam, M., Muhammad, A., & Muhammad, I. R., (2017). Fungal biomass protein production from *Trichoderma harzianum* Using Rice Polishing. *Biomed. Res. Int., 9.*
12. Sarwar, M., (2015). Microbial insecticides - an ecofriendly effective line of attack for insect pests management. *Int. J. Eng. and Adv. Research Tech., 1,* 4–9.
13. Gergerich, R. C., & Dolja, V. V., (2006). Introduction to plant viruses, the invisible foe. *Plant Health Instr.*
14. Gomez, P. L., & Robinson, J. M., (2018). Vaccine manufacturing. In: Stanley, A. P., Orenstein, W. A., & Edwards, K. M., (eds.), *Plotkin's Vaccines* (pp. 51–60. e1). Elsevier.
15. Borkar, T. G., & Goenka, V., (2019). Techniques employed in production of traditional vaccines commonly used by military forces: A review. *J. Arch. Mil. Med., 7,* e96149.
16. Badger-Emeka, L., (2013). Production of antibiotics. In: Ogbonna, U., & Enweani, (eds.), *Fundamentals, Industrial and Medical Biotechnology* (pp. 148–167). Universal Academic Services: Beijing.
17. Rahman, S. U., Rasool, M. H., & Rafi, M., (2012). Penicillin production by wild isolates of *Penicillium chrysogenum* in Pakistan. *Braz. J. Microbiol, 43,* 476–481.

18. Sarris, D., & Papanikolaou, S., (2015). Biotechnological production of ethanol: Biochemistry, processes and technologies. *Eng. Life Sci., 16*, 307–329.

19. Carrillo-Nieves, D., Ruiz, H. A., Aguilar, C. N., Ilyina, A., Parra-Saldivar, R., Torres, J. A., & Martínez-Hernández, J. L., (2017). Process alternative for bioethanol production from mango steam bark residues. *Bioresour. Technol., 239*, 430–436.

20. Ouattara, A., Somda, K. M., Cheik, O. A. T., Traore, S. A., & Ouattara, S. A., (2018). Production of acetic acid by acetic acid bacteria using mango juice in Burkina Faso. *Int. J. Biol. Chem. Sci., 12*, 2309–2317.

21. Vidra, A., & Németh, Á., (2018). Bio-produced acetic acid: A review. *Period. Polytech. Chem. Eng., 62*, 245–256.

22. Max, B., Salgado, J. M., Rodríguez, N., Cortés, S., Converti, A., & Domínguez, J. M., (2010). Biotechnological production of citric acid. *Braz. J. Microbiol., 41*, 862–875.

23. Komesu, A., Rocha De, O. J. A., Da Silva, M. L. H., Wolf-Maciel, M. R., & Filho, R. M., (2017). Lactic acid production to purification: A review. *Bio. Res., 12*, 4364–4383.

24. Wee, Y. J., Kim, J. N., & Ryu, H. W., (2006). Biotechnological production of lactic acid its recent applications. *Food Technol. Biotechnol., 44*, 163–172.

25. Negin, A., Khosravi-Darani, K., Mortazavian, A. M., (2017). An overview of biotechnological production of propionic acid: From upstream to downstream processes. *Electron. J. Biotechnol., 28*, 67–75.

26. Ramachandran, S., Fontanille, P., Pandey, A., & Larroche, C., (2006). Gluconic acid: A review. *Food Technol. Biotechnol., 44*, 185–195.

27. Singh, R., Kumar, M., Mittal, A., & Mehta, P. K., (2016). Microbial enzymes: Industrial progress in 21st century. *3 Biotech., 6*, 174.

28. Smitha, M. S., Singh, S., & Singh, R., (2017). Microbial bio transformation: A process for chemical alterations. *J. Bacteriol. Mycol. Open Access, 4*, 47–51.

29. Nassiri, N., & Faramarzi, M., (2015). Recent developments in the fungal transformation of steroids. *Biocatal. Biotransformation, 33*. 1–28.

30. Adrio, J. L., & Demain, A. L., (2003). Fungal biotechnology. *Int. Microbiol., 6*, 191–199.

31. Steensels, J., Snoek, T., Meersman, E., Picca-Nicolino, M., Voordeckers, K., & Verstrepen, K. J., (2014). Improving industrial yeast strains: Exploiting natural and artificial diversity. *FEMS Microbiol. Rev., 38*, 947–995.

CHAPTER 7

Hygiene, Control, and Inspection in Foods

JOSÉ SANDOVAL-CORTES,[1] JOSÉ L. MARTÍNEZ-HERNÁNDEZ,[2] and CRISTÓBAL NOÉ AGUILAR[2]

[1]*Analytical Chemistry Department, School of Chemistry, Autonomous University of Coahuila, Saltillo, México, E-mail: josesandoval@uadec.edu.mx*

[2]*Bioprocesses and Bioproducts Group, Food Research Department, Autonomous University of Coahuila, Saltillo, México*

ABSTRACT

In the last century, hygiene, inspection and control of food have taken an alternative direction since previously it focused on general aspects of microorganisms that cause problems to human health, as well as aspects that would guarantee the contents of food and avoid fraud. Currently, advances in chemistry, microbiology and other areas have allowed important advances, however they have been added to the advances in pathological alterations, some toxicities, and compromised aspects of food safety. For this reason, hygiene, inspection, and sanitary control have had to undergo constant updates, responding to the constants in the diet that can lead to risks to control and that avoid the assurance of the food, in addition to find mechanisms that verify the safety, the value of the products and with it the security in the value nutritious. In this chapter on hygiene, control, and inspection in food approach these aspects.

7.1 INTRODUCTION

Food poisoning are a health problem in the U.S., whit a total of 48 million sick people, 128,000 cases that need hospital care, and 3,000 deaths each year [1]. That is why is important to talk about hygiene and cleaning, here some definitions about it:

1. **Hygiene:** Cleaning and sanitation.
2. **Cleaning:** Removal of organic and mineral residues adhered to surfaces, consisting mainly of proteins, fats, and mineral salts.
3. **Health:** Elimination of pathogenic microorganisms and reduction of the number of saprophytes. Cleaning: reduces the microbial load of surfaces but not to a satisfactory level.

7.2   WASTE REMOVAL

Food employees at delis, bakeries, and snack bars are responsible for the sanitation their work areas, but there are a lot of problems related with that task, one of them is related to the correct use of sanitizers that in some cases are used incorrectly as cleaners [2], then is important to know about the use of different solvents directly related with the kind of food waste, in Table 7.1 there are summarized some solvents or solutions that should be used for waste removal.

TABLE 7.1   Waste Characteristics

Residue	Solubility	Reaction	Heat Modifications
Carbohydrates	Soluble in water	Easy	Caramelization
Fat	Insoluble in water Soluble in alkali Soluble by surfactants	Hard	Polymerization
Proteins	Insoluble in water Soluble in alkali	Hard	Denaturalization
Monovalent mineral salts	Soluble in water Soluble in acids	Hard	Difficult removal
Polyvalent mineral salts	Insoluble in water Soluble in acids	Hard	Difficult removal

1. **Organic Waste:** The main ones are fats and proteins. To remove them, specific chemical transformations:
   i.   **For Fats:** Saponification (soap formation-water soluble) and/ or emulsification (polarization change, making it soluble).
   ii. **For Proteins:** Solubilization.

2. **Mineral Waste:** Complex agents are used for removal and acid solutions are applied.

7.3   WATER QUALITY

By its own physicochemical characteristics and polarity, water can be considered as universal solvent, even more, water is the biological solvent selected by nature, that forces us to use water into the food processes, but water does not come alone, there are a lot of compounds dissolved in water, some of them represents benefits for the health but if certain levels are exceeded, water can be harmful [3]. Here we refer some aspects about water quality.

1. **Physical Aspects:** The presence of ferric ions can stain materials and affect industrial processes.
   i. **Turbidity:** Suspension of materials of any nature such as lama and sand.
   ii. **Flavors and Odors:** Presence of sulfuric acid, methane, CO_2, organic materials, and mineral substances, such as unwanted salts.
2. **Chemical Aspects:**
   1. **Hardness:** A combination of food residues, detergent residues and hard water salts may occur, forming "stones" on the surface of the equipment, where microorganisms can develop (Table 7.2):

TABLE 7.2   Hardness: Presence of Calcium and Magnesium Salts Leached by Water

Hardness	Quality of $CaCO_3$ (mg/L)
Soft water	50
Moderately hard water	50–150
Hard water	150–300
Very hard water	Greater than 300

2. **Acidity and Alkalinity:** Total acidity represents the concentration of free CO_2, mineral, and organic acids, and salts of strong acids, they are corrosive to equipment.
   Alkalinity represents the concentration of carbonates, bicarbonates, CaOH, Mg(OH), Fe(OH), Mn(OH).

3. **Silica:** The presence of SiO_2 is hard to remove. Normal level is about 5–50 mg/L.
4. **Gases:** CO_2 and O_2 are corrosive.
5. **Iron and Magnesium:** They cause the formation of deposits and scabs, color the products (Table 7.3).

TABLE 7.3   Normal Parameters for Drinking Water

Parameter	Concentration	Parameter	Concentration
Turbidity	5 mg/L (max)	Fluorates	1 mg/L
Odor	None	Silicates	Absent
Color	20 (PT/L-Hazen)	Sulfates	250 mg/L
pH	6.5–8.5	Total solids	1,000 mg/L
Total hardness	200 mg/L	Copper	3 mg/L
Total acidity	5–20 mg/L	Lead	0.1 mg/L
Total alkalinity	10–50 mg/L	Iron	0.3 mg/L
Caustic alkalinity	Absent	Magnesium	0.1 mg/L
Oxygen consumed	2 mg/L	Zinc	5 mg/L
Nitrites	Absent	Phenol	0.001 mg/L
Chlorates	200 mg/L	Aerobic pollutants	100 UFC/100 ml
Residual bleach	0.2 mg/L	Mesophiles	–
		Total Coliforms (NMP)	Absent in 100 ml

7.4   MAIN TYPES OF SURFACES USED IN THE FOOD INDUSTRY

Food contact surfaces can play an important role in microbial contamination, surfaces like tables, cutting bords, knives, processing machinery and containers can act as vehicles of pathogens, those surfaces can be made in different materials [4, 21]. Here some kinds of surfaces and its own way to be sanitized or cleaned:

1. **Wood and Rubber:** Hard to sanitize.
2. **Steel, Carbon:** Use neutral detergent.
3. **Tin:** It should not encounter food.
4. **Concrete:** Damaged by acidic foods and cleaning agents.
5. **Ink:** Some are suitable in the food industry.
6. **Stainless Steel:** Corrosion resistant, easy to clean, it is expensive.
7. **Glass:** Water and detergent.

7.5 TYPES AND LEVELS OF MICROBIOLOGICAL CONTAMINATION

Around 200 diseases can be linked to man by food consumption and food-borne diseases are growing around the world, World Health Organization considers that 1 in 10 people suffered a disease related to food consumption in 2010 [5]. The etiological agents can be bacteria, fungi, viruses, parasites, chemical agents and toxic substances of animal and plant origin.

Bacteria are responsible for 70% and 90% of cases, 50% occur in community services (restaurants), 5% in industrialized food, 15% in residences, in home food production, 30% is of unknown origin.

7.6 WATER BACTERIOLOGY

Drinking water and water used for industrial purposes, may or may not come from the same origin. They must comply [6] with Public Health regulations (it should not contain coliform bacteria in an amount that contaminates wastewater). In the case of water for industrial use, it must obey more requirements depending on the food produced. Sometimes certain waters that are considered potable are not suitable for use in certain foods:

- In the manufacture of butter and cottage cheese, water containing psychrophilic microorganisms such as *Pseudomonas* and *Alcaligenes* is inadequate due to their capacity for lipolytic activity, which can cause oxidation of the products.
- Hard water should not be used in the manufacture of beer, or in leguminous sausages, as it causes abnormal odors (those responsible are iron and magnesium).
- Water supply: they cause problems in the industries and mucous growth of iron bacteria.

Waste from food factories ordinarily contains a wide variety of organic products, some easily oxidizable and others complex and difficult to decompose. The concentration of waste or "cloacal water" that contains organic matter, is expressed in terms of BOD (biochemical oxygen demand). The amount of O_2 used by aerobic microorganisms and reducing compounds in the stabilization of decomposing matter for a certain time at a certain temperature. Normally a period of 5 days at 20°C is used and the result is expressed in BOD of 5 days: ppm BOD of 5 days per gallons of waste $\times$ 8.34/1,000 = pounds of BOD. When there are residues in the water with high concentrations of BOD, or any high organic matter, quickly they are consumed of 7–8 ppm of O_2 (normal water) for example, residues of the dairy industry, around BOD of 500–2,000 ppm.

7.7 TYPES OF TREATMENT

7.7.1 CHEMICAL TREATMENT

Chemical substances are added for the water sanitation [7], in order to eliminate contaminants by sedimentation or flocculation and to sediment along with a large part of the suspended colloidal materials, including bacteria. Soluble salts of aluminum and iron are the most used.

7.7.2 BIOLOGICAL TREATMENT AND WASTE DISPOSAL

Biological treatments can be applied for the cleaning water [8, 9]:

- Dilution of wastewater with running water;
- Irrigation. When wastewater extends over a considerable area of open textured soil;
- Artificial lakes with or without other treatments;
- Filtration with prepared rock filters;
- Activated waste: Inoculate the expulsion waters with waste from a previous batch by actively airing in suitable tanks;
- Anaerobic tanks: Sedimentation, hydrolysis, putrefaction, and fermentation, followed by aerobic treatment.

7.8 TYPES OF WASTE FROM THE FOOD INDUSTRY

Food industry can generate important environmental impact [10, 11] due to the residues generated during food processing:

1. **Dairy Industry:** High protein content, large amount of lactose and many microorganisms; under anaerobic conditions they are acidified, and their treatment is very difficult (500–2,000 BOD).

2. **Fruit Canning Factory:** Acid wastes (200–2,100 BOD).

3. **Manufacture of Malt, Beer, Distillates, Canned Corn, and By-Products:** Rich in carbohydrates and susceptible to acidification under anaerobic conditions (420–1,200 BOD).

4. **Protein-Rich Wastes:** They can decompose under anaerobic conditions (380–4,700 BOD).

5. **Paper Mill:** The water contains sulfites that hinder its decomposition by microorganisms.

7.9   MICROBIOLOGY OF FOOD PRODUCTS

Minimizing microbial contamination [12, 13] and preserving product quality requires an examination of:

- Raw materials used;
- Cleaning procedures;
- Hygiene and sanitation of equipment;
- Control of the conservation mechanism;
- Supervision of packaging and storage processes.

The correct manufacturing practices (GMPs) are specified by three international organizations:

- Department of Health Education and Welfare;
- Public Health Service;
- Food and Drug Administration (FDA).

They are found in the Federal Regularization Code, title 21, code 128.1 to 128.10.

There are sections dependent on the United Nations that deal with international food trade such as:

1. **FAO:** Food production through the most appropriate systems, in relation to the production, processing, conservation, and distribution itself.

2. **WHO:** More associated with consumer health and with obtaining food with the necessary sanitary quality.

3. UNICEF.

7.10 MICROBIOLOGICAL CRITERIA THAT FOOD MUST MEET

The fundamental objective pursued in establishing the bacteriological parameters of food and ensuring [14–16]:

- Its acceptability regarding public health (which determines infectious diseases and food poisoning);
- That the food is of a satisfactory quality (good quality raw material that will not be damaged throughout the process);
- That it has a pleasant appearance (free of dirt due to fecal matter, mycelia, parasites, etc.;
- That the conservation capacity is normal for the food in question.

7.10.1 DEFINITIONS OF MICROBIOLOGICAL CRITERIA

According to the criteria responsible for the qualities of food, patterns, limits, and specifications are distinguished:

1. **Microbiological Pattern:** The official administrative regulation establishes the maximum number of microorganisms, determined by officially stipulated methods.
2. **Recommended Microbiological Limit:** The maximum tolerable limit of microorganisms, determined by stipulated methods suggested as acceptable for a given food.
3. **Microbiological Specification:** The maximum tolerable limit of microorganisms, determined by stipulated methods of internal use in a way to control the quality of your product.

7.11 GENERAL MICROBIOLOGICAL METHODS

Depending on the food sample, one or more methods can be chosen that can evaluate the microbiological quality of the food, that is, perform the general microbiological analysis, or the specific pathogen search. It is important to mention that in the general analysis you can include direct microscopic count, plate count or determination of the most probable number.

7.11.1   GROUPS OR SPECIES OF IMPORTANT MICROORGANISMS IN THE EVALUATION OF FOOD MICROBIOLOGICAL QUALITY

Its objective is to obtain information on the quality of the hygienic-sanitary and public health aspects, as well as the probable shelf life of the food. They are grouped according to the ICMSF:

- (g.a) Microorganisms that are not a direct health risk. They cause food spoilage or reduce their half-life. They are not pathogenic and offer general information.
- (g.b) Microorganisms that have a low and indirect rise to health. They are indicator microorganisms that provide information on hygienic-sanitary conditions in the processing and adaptation of preservation techniques.
- (g.c) Microorganisms that present a direct risk, moderate, and with limited diffusion, are potentially pathogenic bacteria, responsible for food poisoning and infections.
- (g.d) Microorganisms that present a direct, moderate risk with potentially widespread diffusion, are potentially pathogenic bacteria; pathological processes occur as when they are ingested in smaller doses, being able to be easily diffused by food, from a contaminant. The microorganisms must be absent in a 25-gr serving of food.
- (g.e) Microorganisms that present a direct and serious risk, are highly pathogenic, whose presence is absolutely prohibit in food for human consumption.

7.12   RISK ANALYSIS AND CONTROL OF CRITICAL POINTS: HACCP

The preventive system of control and microbiological risks, through careful analysis of the ingredients, products, and processes for determining the components or areas that must be maintained under strict control to ensure that the final product has the microbiological specifications established for it.

HACCP allows to identify the critical stages to ensure the product and where to concentrate technical resources to ensure that critical operations are under control [17–19]. The comprehensive method applies to all phases of the food production and consumption cycle and should consider the raw material, ingredients, process stages and potential consumer abuse. The objective is to contribute to food security so that a safe and contamination-free food is obtained. The contamination can be of pathogenic, toxic, chemical waste and foreign materials origin.

The main indicators are:

- *Escherichia coli*;
- *Staphylococcus aureus*;
- Total count;
- Total coliforms;
- *Bacillus cereus*;
- *Salmonella*;
- *Clostridium perfringens*;
- Fungi and yeasts.

System Structure: The System Is Composed of Five Elements

1. **Identification and Analysis of Risk Factors and Potential Hazards:** It shows the presence of risks at all stages of the process. The food cycle is analyzed by locating the sources of potential risk and the specific moment of contamination, determining the chances of survival of microorganisms and their permanence and diffusion.

2. **Determination of the Critical Control Points:** It requires knowing the process, establishing its flow chart through a clear and simple description, covering all the manufacturing stages. The flow chart must contain information like raw materials, process stages, packing stages, process conditions (time, temperature), pH, AA, biological, chemical, and physical contamination, inactivation of essential nutrients, formation of undesirable substances.

3. **Selection of Criteria for Control:** Specify these criteria and determine critical limits. The criteria are the specifications and physical, chemical, and biological characteristics that must be achieved in the process to guarantee food quality. The most common microbiological criteria are total contamination of mesophiles, total, and fecal coliforms, filamentous fungi and yeasts, detection of *Salmonella*. Their limits must be set based on guidelines and patterns of legislation, literature, practical experience, prior data collection and internal standards of each company.

4. **Monitoring of Critical Points and Decision Making when the Results Indicate that the Proposed Microbiological Parameters are being Achieved:** At the end of the monitoring, it must include the process stages, the danger involved, the preventive control measure, the variability involved, the critical limit, the variability, corrective measures, and the registration number of the form.

5. **Verification that the System is Working as Planned:** Establishing corrective actions, setting up an audit registration system. The system is in operation is verified, if the critical control points are appropriate or not, if the monitoring is being practiced.

KEYWORDS

- *Alcaligenes*
- biochemical oxygen demand
- critical control points
- Food and Drug Administration
- microorganisms
- *Pseudomonas*

REFERENCES

1. Machado, R. A. M., & Cutter, C. N., (2017). Sanitation indicators as a tool to evaluate a food safety and sanitation training program for farmstead cheese processors. *Food Control, 78*, 264–269.

2. Crandall, P. G., Bryan, C. A. O., Grinstead, D. A., Das, K., Rose, C., & Shabatura, J. J., (2016). *Role of Ethnographic Research for Assessing Behavior of Employees During Cleaning and Sanitation in Food Preparation Areas, 59*, 849–853.

3. Salari, M., Salami, E., Hosein, S., Ehteshami, M., Oliveri, G., Derakhshan, Z., & Nikbakht, S., (2018). Quality assessment and artificial neural networks modeling for characterization of chemical and physical parameters of potable water. *Food Chem. Toxicol., 118*, 212–219.

4. Falcó, I., Verdeguer, M., Aznar, R., Sánchez, G., & Randazzo, W., (2019). Sanitizing food contact surfaces by the use of essential oils. *Innov. Food Sci. Emerg. Technol., 51*, 220–228.

5. Li, W., Pires, S. M., Liu, Z., Ma, X., Liang, J., Jiang, Y., Chen, J., et al., (2020). Surveillance of foodborne disease outbreaks in China, 2003 – 2017. *Food Control, 118*, 107359.

6. Bigham, T., Dooley, J. S. G., Ternan, N. G., Snelling, W. J., Héctor-Castelán, M. C., & Davis, J., (2019). Assessing microbial water quality: Electroanalytical approaches to the detection of coliforms. *Trends Anal. Chem., 121*, 115670.

7. Jacobs, L., Persia, M. E., Mccoy, J., Ahmad, M., Lyman, J., & Good, L., (2019). Impact of water sanitation on broiler chicken production and welfare parameters. *J. Appl. Poult. Res., 29*, 258–268.

8. Fuss, M., Vergara-Araya, M., Barros, R. T. V., & Poganietz, W. R., (2020). Implementing mechanical biological treatment in an emerging waste management system predominated by waste pickers: A Brazilian case study. *Resour. Conserv. Recycl., 162*, 105031.

9. Recalde, M., Woudstra, T., & Aravind, P. V., (2018). Renewed sanitation technology: A highly efficient faecal-sludge gasification-solid oxide fuel cell power plant. *App Energ., 222*, 515–529.

10. Kosseva, M. R., (2013). Recent European legislation on management of wastes in the food industry. In: Kosseva, M. R., & Webb, C., (eds.), *Food Industry Wastes* (pp. 3–15). Academic Press.

11. Liu, H., Wang, F., & Liu, Y., (2016). Hot-compressed water extraction of polysaccharides from soy hulls. *Food Chem., 202*, 104–109.

12. Banach, J. L., Bokhorst-Van De, V., H. V., Van, O. L. S., Van, D. Z. P. S., Fels-Klerx, H. J. V. D, & Nierop, G. M. N., (2017). The efficacy of chemical sanitizers on the reduction of *Salmonella typhimurium* and *Escherichia coli* affected by bacterial cell history and water quality. *Food Control, 81*, 137–146.

13. Grasso, E. M., Grove, S. F., Halik, L. A., Arritt, F., & Keller, S. E., (2015). Cleaning and sanitation of *Salmonella* contaminated peanut butter processing equipment. *Food Microbiol., 46*, 100–106.

14. Kim, N. H., Lee, N. Y., Kim, M. G., Kim, H. W., Cho, T. J., Joo, I. S., Heo, E. J., & Rhee, M. S., (2018). Microbiological criteria and ecology of commercially available processed cheeses according to the product specification and physicochemical characteristics. *Food Res. Inter., 106*, 468–474.

15. Reich, F., Valero, A., Schill, F., Bungenstock, L., & Klein, G., (2018). Characterization of *Campylobacter* contamination in broilers and assessment of microbiological criteria for the pathogen in broiler slaughterhouses. *Food Control, 87*, 60–69.

16. Zwietering, M. H., Ross, T., & Gorris, L. G. M., (2014). Food safety assurance systems: Microbiological testing, sampling plans, and microbiological criteria. In: Motarjemi, Y., (ed.), *Encyclopedia of Food Safety* (pp. 244–253) Academic Press.

17. Dzwolak, W., (2019). Assessment of HACCP plans in standardized food safety management systems - the case of small-sized polish food businesses. *Food Control, 106*, 106716.

18. Hasnan, N. Z. N., & Mohd, R. S. H., (2020). Modernizing the preparation of the Malaysian mixed rice dish (MRD) with cook-chill central kitchen and implementation of HACCP. *Int. J. Gastron. Food Sci., 19*, 100193.

19. Vukman, D., Viličnik, P., Vahčić, N., Lasić, D., Niseteo, T., Panjkota, K. I., Marković, K., & Bituh, M., (2021). Design and evaluation of an HACCP gluten-free protocol in a children's hospital. *Food Control, 120*, 107527.

20. Brasil, C., Barin, J. S., Jacob-Lopes, E., Menezes, C. R., Zepka, L. Q., Wagner, R., Campagnol, P., & Cichoski, A. J., (2017). Single step non-thermal cleaning/sanitation of knives used in meat industry with ultrasound. *Food Res. Inter., 91*, 133–139.

21. Múgica, V. R., Ramírez, A. C., Muro, F. I., Santamaría, O. C., & Alba, E. F. (2018). New methods for biofilms control in the food industry. *22nd International Congress on Project Management and Engineering*, pp. 925–936.

CHAPTER 8

Molecular Methods for Microorganism Detection

ADRIANA C. FLORES-GALLEGOS, RAÚL RODRÍGUEZ-HERRERA, and CRISTÓBAL NOÉ AGUILAR

Bioprocesses and Bioproducts Research Group, Food Research Department, Autonomous University of Coahuila, Saltillo, México, E-mail: carolinaflores@uadec.edu.mx (Adriana C. Flores-Gallegos)

ABSTRACT

Assessment of microbial diversity associated with a fermented food is paramount for both producers and consumers. However, it is generally difficult to cultivate most bacteria or stressed cells in food using classical microbiology techniques. This has led to the introduction of molecular methods in recent years that allow rapid identification of microorganisms from pure cultures of the fermented food or drink without need of isolation or enhanced differential growth. These methods allow to obtain the fingerprint of a microbial community based on the DNA/RNA total of a community. Among these techniques, the polymerase chain reaction (PCR) is an essential tool as rapid, sensitive, inexpensive and reproducible. Using the 16S, 18S and 23S RNAr gene and molecular markers as AFLP, RFLP and RAPD it has been possible to identify bacteria, yeasts and fungi in food and starter cultures. It has been also possible to monitor starter cultures at any stage of the fermentation process, the selection of mutants and starter cultures with genetically engineered strains with greater capacities than those found in nature and the detection and identification of microbial pathogens, ensuring quality microbiological food.

8.1   INTRODUCTION

Microorganisms play a very important role in human life since they are not only pathogenic to animals, plants or humans, but many of them are beneficial since they can be used to control other harmful microorganisms, insects, animals or plants, for humans or their economy, they can also help improve the living standards of human beings as in the case of the prebiotics and bacteria used in yogurt, they can also be used for the production of food for humans or animals, such as case of edible mushrooms (Mushrooms, Pleurotus, Huitlacoche, etc.), or in the fermentation of foods such as cheese, wine, bread, etc. They can be used for environmental rehabilitation (hydrocarbon-degrading bacteria, wastewater treatments, etc.). In addition, microorganisms can be used to produce metabolites of economic importance (antibiotics, pigments, amino acids, hormones, etc.), also, microorganisms help in the proper development of crops (Nitrogen-fixing bacteria, mycorrhizae).

In epidemiology, the identification not only of the microorganism in question is very important, but often identification strains help to determine the cause of an infectious outbreak, detect the cross-transmission of pathogens, determine the source of infection, recognize strains particularly virulent organisms, and monitor vaccination programs [14]. In the clinical area and in agriculture, the basis for an effective treatment and cure of a patient, animal or plant is the rapid diagnosis of the disease and its causal agent [19]. This is especially important when trying to diagnose slow-growing or difficult-to-treat microorganisms such as Chlamydia, mycobacteria, mycoplasmas, Herpes viruses, and enteroviruses [13]. In ecological research, a rapid and correct identification of microorganisms helps to quantify organisms and to characterize their activity in different habitats [8]. The objective of this chapter is to emphasize the importance of molecular techniques in the identification of microorganisms, describe some of the most widely used molecular techniques, as well as some techniques with the greatest potential for the molecular diagnosis of microorganisms in the near future and relate the above to some achievements in our laboratory.

8.2   METHODS OF IDENTIFICATION OF MICROORGANISMS

Given the great importance of microorganisms for humans, a series of methodologies have been developed to differentiate one microorganism from another, which have traditionally been based on the observation of the symptoms that the hosts present, on the macro or microscopic observation of the microorganism or on the reproductive structures of the microorganisms, of the development of the pathogen in a specific culture medium, of the staining produced by applying some dyes on the infected tissue, or on the pathogen and the reaction of an antibody to the presence of a pathogen.

The identification of a microorganism (pathogen) based on the symptoms presents several disadvantages, for example, the symptoms are different and numerous; they depend on the host of the pathogen, and even on the factors of the environment in which the pathological process develops. The identification of pathogens based only on symptoms can give rise to serious errors in the diagnosis of a disease, since two different pathogens can produce the same symptoms in the same host, the same pathogen can produce different symptoms in different hosts and even the symptoms can vary depending on the environment where the pathological process develops.

Detection of a microorganism based on macro or microscopic observation is more efficient than identification based only on symptoms. Observation at the macro or microscopic level is a simple procedure, it is a direct detection of pathogens and can differentiate organisms morphologically. It has the following disadvantages: it is a slow, laborious, and tedious procedure since the microorganism has to be isolated, which is not possible in all cases and it must be grown in a culture medium until it reaches a size or stage of adequate growth for identification. Identification in the microscope is only possible in very specific cases. It is of low sensitivity, it requires very specialized equipment for the identification at the microscopic level of some microorganisms (viruses, mycoplasmas, viroid, prions). The structure of a microorganism can change due to the influence of the environment, and it takes a lot of experience to differentiate one species from another at a microscopic level.

The identification of pathogens by staining or the development of the pathogen in a specific culture medium, can detect a wide range of variants, does not require very sophisticated equipment, and are easy tests to perform. However, it has the disadvantages that the time to identify a microorganism can be very long (in some cases up to 25 days), in addition a large number of tests are required to be sure of the identification of the pathogen (a single test is not sufficient in most cases). Because different tests are used, many reagents are needed, which increases the cost of the test significantly. Other procedures, such as in vitro culture and inoculation in mice, are expensive techniques and pathogens can present differences in infeasibility depending on the host.

The use of antibodies for the identification of microorganisms has been used because it is a simple and fast technique. It can be automated, and it can be used to work with several samples at the same time. However, detection cannot be accurate when there are minimal amounts of the microorganism; in the case of pathogens, it does not distinguish between active and latent infections. In addition, antibodies, and antigens for many microorganisms are lacking, in some cases the detection is not precise due to the null specificity of some antibodies, other disadvantages are that the antibodies have a high cost or in some cases they can offer false positives, such as in the first generation of tests for HIV detection, which were based on pure viral antigen preparations derived from the virus culture, and false positives were obtained, due to the presence of residual cellular antigens which were incorporated into the viral particle during ripening [19]. Molecular techniques are based on the analysis of the nucleic acids extracted from the microorganisms either directly or from a sample containing the microorganism in question.

8.3   ADVANTAGES AND DISADVANTAGES OF MOLECULAR METHODS

Molecular techniques for the identification of microorganisms are currently booming due to their specificity (they can detect only the target molecule or microorganism), sensitivity (they are capable of detecting the presence of a single microorganism), they are fast (less than 24 hours), and are capable of automation (they allow a diagnosis in less time and reduce costs). The use of molecular techniques has allowed the identification of new microorganisms, which had not been possible to cultivate and identify by traditional techniques [8]. Furthermore, they allow the study of microbial populations without isolating, therefore, the biases that can arise with the cultivation of microorganisms are avoided [3]. Likewise, these techniques allow the detection of highly pathogenic microorganisms, which can be identified dead, thus avoiding the risks of infection of the analyst [9]. Molecular techniques have proven to be very useful in clinical situations where conventional methods are very insensitive (e.g., during the asymptomatic stage of HIV infections), very slow (mycobacterial culture) or very complicated to be used on a large scale (e.g., viral isolation). Another important application is in monitoring the emergence of genome mutations, e.g., selection of resistant variants during antiviral/antibiotic therapy [19].

Some of the disadvantages associated with molecular techniques are: (i) do not distinguish between living and dead organisms, although sometimes this is not exactly a disadvantage in the case of *Staphylococcus aureus*, the bacteria dies at temperatures that do not degrade the toxin, thus detecting dead bacteria is indicative of the presence of toxins in the sample; (ii) knowledge of specific sequences must be available for each pathogen; (iii) trained personnel are required to carry out the identification tests; (iv) sophisticated equipment is required which increases the initial investment; and (v) procedures with multiple stages which increases the possibility of errors, in addition to the possibility of obtaining false positives and false negatives. As the use of these molecular techniques increases, costs will tend to decrease, on the other hand, the problem of false positives and false negatives can be solved by including a positive control and a negative control in each analysis [1].

8.4   MOLECULAR DETECTION OF MICROORGANISMS

A wide variety of molecular techniques have been proposed for the detection of microorganisms. Describing each of the proposed techniques is beyond the scope of this book. Most of the molecular techniques are based on hybridization of nucleic acids or polymerase chain reaction (PCR). One of the simplest molecular techniques to identify bacteria consists in determining the percentage of guanine and cytosine present in the bacterial DNA, in this case the extraction of DNA from a pure colony is carried out and the percentage of guanine and cytosine is determined by spectrometry, for example, it has been determined that the genus Pantoea has a percentage of guanine and cytosine that goes from 50 to 58% and bacteria of the genus Pseudomona has a percentage of guanine and cytosine that goes from 58 to 70%. However, this technique has the disadvantage of not being able to identify species, bacteria could only be identified at the genus level, in addition, some genres of bacteria have very similar percentages of guanine and cytosine, therefore the possibilities of obtaining false positives (identifying an organism present when in fact it is absent) increase.

8.5 TECHNIQUES BASED ON DNA HYBRIDIZATION

These techniques were developed from the knowledge that two single nucleic acid chains having complementary sequences to each other can join or hybridize to form a double chain. The single strands can be of DNA or RNA, or one of DNA and the other of RNA.

8.5.1 RESTRICTION FRAGMENT LENGTH POLYMORPHISM (RFLP)

This technique is based on hybridization and consists of extracting DNA from a pure culture of a microorganism or from an infected sample by digesting the DNA with restriction enzymes (enzymes that cut DNA at specific sites). These DNA fragments are subsequently separated by electrophoresis in an agarose gel. DNA is negatively charged, if placed in an electric field close to the negative pole it would migrate towards the positive pole through the gel or matrix, the friction between the matrix and the fragment causes small fragments to migrate faster than large fragments, subsequently the DNA separated in the gel is transferred to a solid support, usually nylon or nitrocellulose membranes, and then all the DNA on the membrane is hybridized with a specific DNA probe (fragment single-stranded DNA (150 to 300 nucleotides) of the microorganism that we want to identify. The probe will join the DNA fragment that has its complement. For the detection, the probe must be labeled radioactively (P^{32}) or non-radioactively (digoxigenin). The probe can be isolated from the nuclear genome, the mitochondrial genome, or DNA that codes for proteins (cDNA). This methodology is very specific; however, it is not very fast and it is not very sensitive since it requires 10^3 to 10^6 copies of the target molecule or sequence to give a reliable result, in addition to using radioactivity, which if not properly used and under special laboratory conditions, there may be health risks for the analyst, contamination of the facilities or the environment.

A variant of RFLP's has been designed for the detection of *Listeria*. This consists of fixing a complementary DNA probe to a region of the bacterial ribosomal RNA in a microtube. The complete RNA of the sample is extracted and placed in the microtube with the probe, which will bind to the ribosomal RNA, then another complementary DNA probe is added to another region of the ribosomal RNA which is marked with fluorescence, then the presence of the microorganism by an antibody with anti-fluorescence [6].

8.5.2   SQUASH BLOT TRANSFER

In this technique, the sample to be analyzed is crushed against a solid support, creating a fingerprint, the sample is fixed using alcohol or exposing it to UV. It hybridizes with the specific probe for the microorganism in question, it has the advantage of being a quick technique that can be used in the field or outside the laboratory. However, it is a dirty technique which sometimes does not allow distinguishing a nucleic acid hybridization with which the probabilities of obtaining false negatives (diagnosing the micro-organism as absent when in fact it is present) increased. Another technique is squash blot-like dot blot and RFLP-like Northern blot. However, what is separated is RNA and hybridizes with DNA probes and has advantages and disadvantages like RFLPs.

8.5.3   DOUBLE STRANDED RNA

This technique is mainly used to detect cells infected with a virus that has RNA (retroviruses) as its genetic material. Since the RNA in a cell is single-stranded and the presence of double-stranded RNA implies replication of the virus within the cell, it is a relatively simple technique since it includes few steps such as extraction and purification of the double strand RNA and its separation in polyacrylamide gels, however, has the disadvantage of identifying only cells infected with retroviruses without being able to specify the taxonomy of the virus, therefore it is a very limited technique.

8.5.4   PCR-BASED TECHNIQUES

Polymerase chain reaction (PCR) or in vitro enzymatic amplification of a specific DNA segment of the target microorganism. This is the most used technique for the detection of microorganisms, it is mainly based on making millions of copies of a specific DNA segment of a microorganism. In this technique, the double strand of DNA is separated into two single strands, subjecting it to higher temperatures at 92°C then two small complementary DNA segments hybridize or align one to one strand and the other to the other. These small segments are called primers, which must be designed in such a way that they only hybridize with the DNA of the desired microorganism and their size generally ranges from 18 to 25 bases. Once the primers are attached to the complementary chains, a DNA polymerase is added to the reaction, which will form the new chain, taking as its base the single chain and one of the ends of the initiator (3'). After this first cycle, if we started with a single double chain of DNA, we would have two double chains. This amplification increases geometrically as the number of cycles is repeated, so after two cycles we would have four double DNA chains, and after three, eight, and, after 30 cycles we would have a large amount of amplified DNA to be visualized with the naked eye. The advantages of PCR over other detection techniques have been reported on several occasions. Detection of *C. trachomatis* by PCR turned out to be more accurate, with a sensitivity of 90–100% and specificity greater than 97% than detection by ELISA [13]. Likewise, PCR has helped to identify infectious microorganisms related to diseases previously identified with a non-infectious origin.

The PCR technique has been used to identify viruses, fungal bacteria, phytoplasmas, nematodes, etc., an example is the method developed by Ayala et al. [1] for the identification of the bacterium *Clavibacter michiganensis* subsp. *nebraskensis*, which is the cause of the disease known as corn bacterial wilt, in this case, specific primers were designed for a segment between the 16S and 23S rRNA genes, which were compared with the DNA sequences reported in the gene banks to verify that they would only align to a sequence of *Clavibacter michiganensis* subsp. *nebraskensis*. The precision and reproducibility of PCR assays depend on the experience of the analyst. The specificity of the test may be affected by contamination of the sample during sample processing, if the initiators are not specific, or if the PCR conditions are not optimal, allowing non-specific products to be amplified [13], the above can guide the detection of false negatives, in the same way the contamination of the sample with inhibitors (humic acids, polyphenols, carbohydrates, ethanol, etc.), of the polymerase enzyme can also cause false negatives. However, this can be solved by always incorporating positive controls. Contamination of the sample with other DNA or RNA can lead to detection of false positives, this is solved by always adding a negative control [1]. The identification of microorganisms directly from food using PCR presents some challenges such as the low sensitivity caused by the inhibition due to the food matrix, this can be partially restored with an enrichment stage, but the time for diagnosis is lengthened [7]. Even with these disadvantages, the flexibility, automation, speed, and reliability of PCR-based molecular techniques are the most widely accepted molecular techniques today for the detection of microorganisms in general.

8.5.5   RT-PCR

Amplification of a segment of DNA previously formed from RNA. This is one of the most used techniques to identify microorganisms whose genetic material is RNA (retroviruses and viroids). In this case, RNA is first extracted from the microorganism in question or from a sample presumably infected by one of these microorganisms, then the RNA is subjected to a temperature generally of 80°C to linearize RNA since it is generally a single strand and tends to form secondary and tertiary structures. Subsequently, a complementary initiator is added, and it is placed at a temperature between 37 and 40°C so that the initiator is aligned with the complementary chain, then the reaction is placed at 4°C, thus stopping it. Then an enzyme called reverse transcriptase is placed which is going to take charge of forming DNA taking the single-stranded RNA as a template and one end (3') of the primer when the DNA has already been formed, the PCR reaction is carried out, this technique can be performed in a single step if all the reagents are added in the same tube and the specific temperatures are programed for each stage. An example of a microorganism that can be easily identified using this technique is the *Citrus tristeza virus*. Another advantage of this technique in the detection of microorganisms is that it allows the identification of active infections [13].

8.5.6   NESTED PCR

Enzymatic amplification of an internal DNA segment from a previously amplified DNA segment. Nested PCR is a technique that is mainly used when a microorganism is found in very low quantities or when you want to identify if microorganisms from a certain group exist in a sample and then determine which species of microorganisms are present. This technique consists of amplifying a large segment of DNA (700–2,000 bp) with the PCR technique with a pair of primers and subsequently amplifying with another pair of primers an internal segment of the first amplified fragment, that is, this technique is a double PCR, and therefore it is estimated that it is 1,000 times more efficient than a simple PCR. Nested PCR has been used successfully for the identification of samples suspected of being infected with phytoplasmas, and if the sample is positive for the presence of phytoplasmas. It is amplified a second time to determine the specific type of phytoplasmas present in the sample. An example of the use of this technique is the detection of the phytoplasma producing lethal yellowing of coconut trees.

8.5.7   *RAPDS OR RANDOM AMPLIFIED DNA POLYMORPHISM*

In this case a single initiator PCR reaction is carried out which is 8 to 12 bases in size with arbitrary sequence. In the primer alignment step, low temperatures of 37 to 40°C are used to ensure that the primer binds to different regions of the genome of the microorganism to be identified. In this case, the identification is made only from pure cultures of the microorganism and with a single PCR reaction, several segments are amplified at the same time (Figure 8.1). It has the advantage that prior knowledge of the microorganism is not required to identify it, this technique as a commercial identification method is not widely used due to reproducibility problems, however, it is very useful when it is required to determine differences between two microorganisms highly related, being able to detect these differences even at between strains. Padilla [15] using RAPDs managed to identify DNA differences between three strains of *Aspergillus niger* tannin degrading. Another advantage of RAPDs is in the case where it is possible to identify an amplified DNA fragment found in a strain, this can be sequenced and initiators specific for that strain can be designed from that sequence. The same applies to the design of specific initiators at the species or genus level, this technique is known as SCAR (sequence characterized by amplified region) (Figure 8.2). In addition to the disadvantage of poor reproducibility, the presence of a RAPD band does not allow to distinguish between the homozygous and heterozygous states [2].

A technique like RAPDs is known as DNA amplification fingerprinting (DAF), in this technique 5–10 base primers are used to amplify DNA at random genomic, the amplified bands are separated on polyacrylamide gels. This technique has been used for the study of *Salmonella enterica* strains [2].

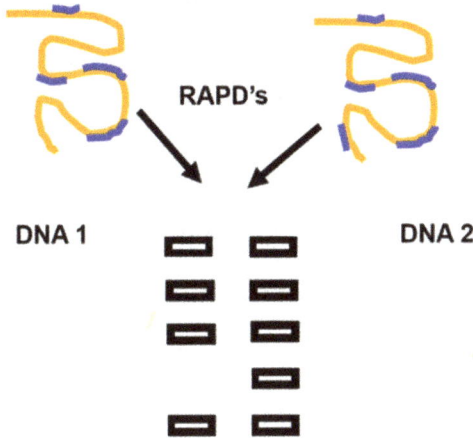

FIGURE 8.1   Random amplified DNA polymorphism.

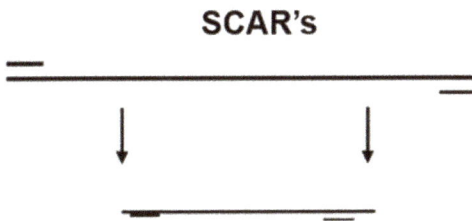

FIGURE 8.2   Sequence characterized by amplified region.

8.5.8   IMMUNO-PCR

Binding of a microorganism to a solid support using an antibody and subsequent DNA amplification by PCR. It is a technique that combines the advantages of immunological techniques (ELISA) and molecular techniques (PCR) so that the combined advantages diminish or eliminate the disadvantages of both techniques. Rocha [22] found that ELISA is inefficient to detect very low levels of the target molecule or microorganism present in the sample to be analyzed, while PCR is inhibited by some components of the sample (phenolic compounds, polysaccharides). These characteristics are combined when it is desired to detect microorganisms directly from plant organs (seeds, tubers, etc.). Rocha in 2003 [22] managed to design an immuno-PCR technique for the detection of *Pantoea stewartii*, directly from the corn seed. In this case, a primary antibody was fixed to a microtube, the excess antibody was washed and the ground seed sample, which was known to be infected with *Pantoea stewartii*, was placed and subsequently left to incubate (1 h) for antigen binding. The antibody was washed to remove the sample and in the tube with the antigen and the antibody the PCR was performed (Figure 8.3).

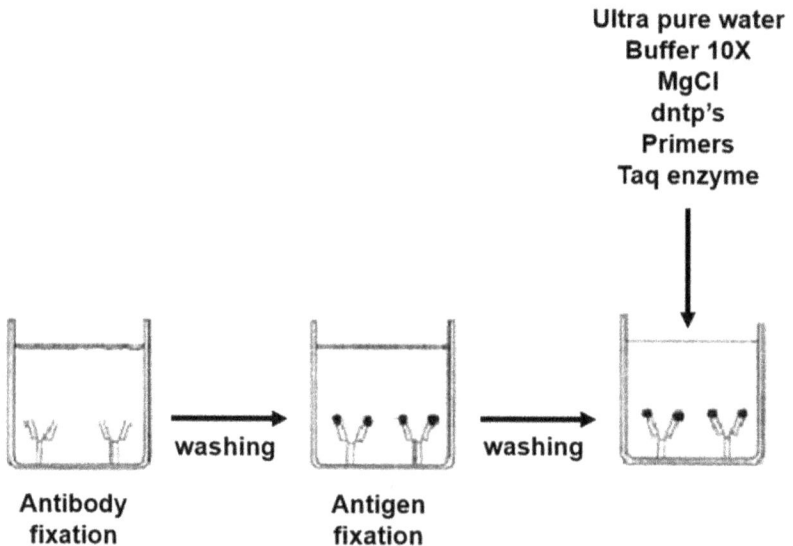

FIGURE 8.3   Immuno-PCR technique steps.

8.5.9   *IMMUNOMAGNETIC PCR*

Like the previous one, in this case the support is a magnetic sphere, which is removed from the sample with a magnet and then the PCR is performed.

In this technique, a primary antibody is attached to metal beads which are placed inside the sample to be analyzed, it is allowed to incubate for antigen-antibody binding, these metal beads are subsequently extracted with a magnet, washed, and placed in a tube in which PCR is performed. This technique has been used for the detection of *Listeria monocytogenes* in food.

8.5.10 PCR-MULTIPLE

One of the problems with the PCR techniques discussed above is that they only allow the identification of a single pathogen at a time. In order to solve this problem, the technique known as multiple-PCR has been designed, allowing the detection of different molecules or target microorganisms in a single reaction. In this technique, a pair of initiators are incorporated into the reaction for each one of the molecules or microorganisms to be detected, which allows to save time and cost for the detection of different microorganisms simultaneously. This technique presents challenges such as designing all the initiators to add to the reaction with a similar alignment temperature, they must not present homology between them, in addition to amplifying fragments of different lengths. The sensitivity of a multiple PCR is on the order of 10^2 cells or 2.9 pg of target DNA. However, this sensitivity can be affected by the inefficient design of the initiators [23]. This technique has been used for the detection of different Listeria species in the same sample, as well as for detecting different microorganisms at the same time, such as *Salmonella*, Campylobacter, Shigella, and *E. coli* [2]. Multiple PCR has also been used to detect different genes from the same organism in a sample. Infected cells do not retain all the genes of the human papilloma virus, therefore using multiple PCR the presence of several viral genes was identified at the same time, which confirmed the causal agent [10]. This technique is also used to identify which antibiotic resistance genes have a certain bacterial strain [16].

8.6 IDENTIFICATION OF HIGHLY SEPARATED MICROORGANISMS

In some occasions, the analyst has to identify a special strain of a micro-organism since this presents outstanding characteristics to produce an enzyme or adaptation to a certain condition, or when it is required to preserve the identity of a particular strain, which would help determine if it was contaminated, also when establishing genetic relationships (phylogenetics) between different strains of a species, different species of a genus, different genera of the same family or between different families of microorganisms. All these cases occur when a strain with a previously unreported characteristic is identified, therefore molecular techniques allow control over the production and dissemination of this strain. Some of the molecular techniques that help to identify highly related microorganisms in addition to those previously discussed are: Rep-PCR, PCR-RFLP, AFLPs, among others.

8.6.1   REP-PCR

This technique is based on the simultaneous amplification of different regions with repeated sequences, it has been used preferably in bacteria, amplifying the BOX, ERIC, and Rep regions. In this case, several fragments of different sizes are obtained which are simultaneously amplified, which allows us to distinguish two highly related microorganisms when comparing the amplification patterns of these microorganisms. It has been proposed that the Rep-PCR technique can be used to elucidate the genetic bases of phenotypic variability between different species, since reproducible banding patterns of similar complexity are produced and allow the differentiation of strains at the subspecies level [9]. This technique has been used for the identification of different species of *Xanthomonas* and *Francisella*.

8.6.2   PCR-RFLP

This technique amplifies by PCR a specific segment of a microorganism and subsequently digest the amplified DNA with restriction enzymes, this allows to determine differences between the patterns of amplified bands or between the patterns of the fragments resulting from DNA digestion. There are different variants of this technique depending on the specific fragment amplified and digested. Between the most commonly amplified and subsequently digested segments are the ribosomal genes (16S in the case of prokaryotes and 18S in the case of eukaryotes), the spaces between the set of ribosomal genes (IGS intergenic spacer) (Figure 8.4) and the internal transcribed spacer (ITS) (Figure 8.5) of the ribosomal gene set. The 16S ribosomal gene is highly conserved but with enough variability to make distinctions between genera and species [8]. Initiators have been developed from the 16S ribosomal gene sequence within a very wide range (universal), from the domain level (bacteria, archaea, or eukaryotes), groups (methanogenic bacteria, sulfate-reducing bacteria, lactic bacteria, etc.), to individual genres (*Escherichia*) [8]. However, when the resolution of the analysis based on the 16S rRNA gene is too small to make inferences about the relationship of highly related microorganisms, the use of the gyrB gene has been proposed, which encodes the B subunit of the DNA gyrase polypeptide. It has been estimated that this gene evolves faster than the 16S rRNA gene while still maintaining a high correlation with the total genome homology analyzed by total DNA-DNA hybridization and microarrays [23]. Methodologies have been established for the rapid identification based on PCR-RFLP of four highly related species of *Carnobacterium* which are important in the food industry [2]. Amplification of the IGS segments and their subsequent digestion with the Hind III enzyme allowed different strains of *Penicillum purpurogenum* to be identified [5]. Amplification of the ITS segments and their subsequent digestion with the Xho I enzyme allowed the identification of different highly related strains from *Aspergillus niger* [15].

Intergenic spacer (IGS)

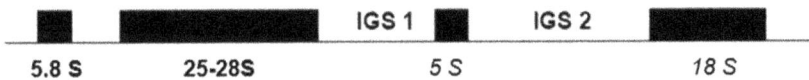

FIGURE 8.4   Location of IGS in DNA.

Internal transcribed spacer (ITS)

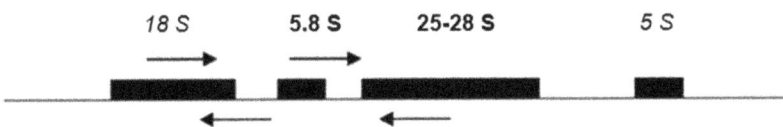

FIGURE 8.5   Location of ITS in DNA.

8.6.3   AMPLIFIED FRAGMENT LENGTH POLYMORPHISMS (AFLPS)

This technique is based on the amplification of a subset of DNA fragments generated by digestion with restriction enzymes. The DNA of the microorganisms is isolated, purified, and digested with the restriction enzymes Eco RI and MseI. The restriction fragments are then ligated to adapters which contain restriction sites at each end of the fragment and a sequence homologous to the primer mating site. The primers used for amplification contain DNA sequences homologous to the adapters and contain from one to two selective bases to the 3' ends of the fragments. Thus, selective nucleotides allow the amplification of only a subgroup of the cut fragments. AFLPs have been shown to be reproducible, and a good way to differentiate highly related strains of microorganisms. Commercial detection is restricted by the high cost of the sequencer used to visualize the amplified fragments [14]. Restrepo et al. [21] evaluated the efficiency of the AFLPs technique to detect genetic variation between different *Xanthomonas axonopodis* strains.

8.6.4   DENATURING GRADIENT GEL ELECTROPHORESIS (DGGE)

This technique can separate highly related sequences due to their different mobility in denaturing gradients [12]. In this technique, some of the genes that code for ribosomal RNA are commonly amplified (16S rDNA, 18S rDNA). The denaturing environment is created by maintaining a uniform temperature (50–60°C) and a linear gradient with urea and formamide. Increasing a denaturing concentration separates the double strands of DNA into different domains. When a domain reaches its denaturation temperature at a position on the gel, the helix partially denatures, and the migration of the molecule practically stops. Variations in the sequence of the domains cause their denaturation temperatures to be different; therefore, they migrate to different positions within the denaturing gradient, effectively separating the domains [8]. In 1997, Kowalchuk et al. [12] amplified the 18S rRNA gene from 20 fungal species and subjected the amplified products to DGGE, they also sequenced the separated bands with DGGE, with this it was possible to identify some fungal strains.

8.6.5 TEMPERATURE GRADIENT GEL ELECTROPHORESIS (TGGE)

Similar to DGGE, in this case the gradient is created by gradually increasing the temperature during electrophoresis. The denaturing environment is formed by a constant concentration of urea in the gel, combined with the temporal temperature gradient [8]. This technique can be used to study the diversity of species in a community of microorganisms and thus compare the diversity of species between communities. This comparison is based on the fact that the number, the precise position, and the intensity of the bands reflect the number and abundance of the dominant types of rDNA in a sample and thus two communities of microorganisms can be compared [4].

8.7 FUTURE OF THE DETECTION OF MICROORGANISMS

Currently there are very promising techniques that in the coming years will undoubtedly play an important role in the detection and quantification of a microorganism in a sample, the most promising are discussed in subsections.

8.7.1 REAL-TIME PCR (RT-PCR)

In this technique, in addition to the traditional components of a normal PCR, a DNA probe is added with a quencher at the end and at the other end a beacon, this probe is complementary to the DNA that you want to amplify, therefore it will align with all the DNA complement chains, when the DNA polymerase amplifies the fragments release this probe from the DNA to which it is attached, when released, fluorescence is emitted and can be detected, so the more DNA is amplified the more light will be emitted and the number of amplified DNA fragments can be estimated. With this technique we can quantify the number of DNA molecules of the microorganism present in a sample. This technique has been used to quantitatively measure viral load in order to monitor the response to a therapy by patients with HIV, cytomegalovirus, or Hepatitis C virus infections [13]. The quantification of a certain microorganism in a sample is of prognostic importance, helps to predict the progress of a disease, and is used to assist in the decision of medical treatment. Furthermore, the real-time PCR (RT-PCR) technique is adapted for the evaluation of many samples at the same time, allowing a fast, reliable, efficient, and cheap laboratory diagnosis [11].

8.7.2 MICROARRAYS OR MICROCHIPS

This technique uses the complementarity of the nitrogenous bases of the DNA and consists of an ordered arrangement of hundreds or thousands of DNA sequences (oligonucleotides, cDNA, etc.), deposited on a solid surface (1.2×1.2 cm.). The solid supports are pieces of silicon or glass. For the detection of microorganisms, an array (chip) can be constructed with DNA from different microorganisms, then a sample containing DNA from one or more microorganisms can be deposited and hybridized on the array and any unhybridized DNA can be removed by washing. The array is then analyzed and scanned to determine with which specific DNA the sample's DNA was hybridized and to identify the microorganism(s) present in the sample. This technique, the reagents used and the image digitizer used to scan the microarrays are very expensive [18].

8.7.3 SEQUENCING

Determining the order or sequence of bases through a segment of DNA is one of the central techniques of molecular biology. There are two techniques for sequencing; one based on an enzymatic method and the other on a chemical method. The requirements that a sequence must have in order to be used in the identification of microorganisms are 1) the sequence must be conserved between a relatively large number of organisms, 2) its change rate must be constant over long periods and between different organisms and it must allow inference of the evolutionary distance between a wide range of life forms; the sequence must not be subject to wide discrepant degrees of evolutionary pressure, 3) the sequence must not have been shared between different organisms by horizontal transmission. Finally, the sequence must be feasible, if it is amplified and detected in various ways [20]. Sequences that meet the above requirements are the regions encoding the small 16S ribosomal units in prokaryotes and 18S in eukaryotes. Vinayagamoorthy et al. in 2003 [24] reported a nucleotide sequence-based method that can simultaneously generate known short nucleotide sequences from a few segments of the same genome or from different genomes. Sequencing is one of the methods with the most future since it only amplifies by PCR the 16S or 18S ribosomal gene, to separate the amplified bands, to sequence or to order the found bands, later to compare these sequences with the sequences deposited in the gene banks and being able to identify the microorganism(s) present in a sample.

KEYWORDS

- **amplified region**
- **DNA amplification fingerprinting**
- **internal transcribed spacer**
- **polymerase chain reaction**
- **restriction fragment length polymorphism**
- **retroviruses**

REFERENCES

1. Ayala-Labarrios, L. A., Rodríguez-Herrera, R., Aguilar-González, C. N., Lara-Victoriano, F., & Quero-Carrillo, A. R., (2005). Detection of *Clavibacter michiganensis* subsp. Nebraskensis (Schuster, Hoff, Mandel and Lazar) Vidaver and Mandel, using polymerase chain reaction. *Mex J. Phytopathol., 22*, 239–245.
2. Babalola, O. O., (2003). Molecular techniques: An overview of methods for the detection of bacteria. *Afr. J. Biotechnol., 2*,710–713.
3. Chan, O. C., Wolf, M., Hepperle, D., & Casper, P., (2002). Methanogenic archaeal community in the sediment of an artificial partitioned acidic bog lake. *Microbiol. Ecol., 42*, 119–129.
4. Eichner, C. A., Erb, R. E., Timmis, K. N., & Wagner-Dobler, I., (1999). Thermal gradient gel electrophoresis analysis of bioprotection from pollutant shocks in the activated sludge microbial community. *Appl. Environ. Microbiol., 65*,102–109.
5. Espinosa, H. T. C., (2004). Caracterización morfológica, fisiológica y molecular de tres cepas fúngicas productoras de pigmentos. BSc dissertarion, Universidad Autónoma de Coahuila, Saltillo, Coahuila, México.
6. Giese, J., (2001). New microorganism detection methods introduced. *Food Technol., 55*, 68–69.
7. Grant, K. A., Dickinson, J. H., Payne, M. J., Campbell, S., Collins, M. D., & Kroll, R. G., (1993). Use of polymerase chain reaction and 16S rRNA sequences for the rapid detection of *Brochothrix spp.* in food. *J. Appl. Microbiol., 74*, 260–267.

8. Jan, J., & LeBorgne, S., (2001). Uso de técnicas moleculares para realizar estudios de biodiversidad microbiana en ambientes petroleros. *Biotecnología, 5*, 103–109.

9. Johansson, A., Ibrahim, A., Goransson, I., Eriksson, U., Gurycova, D., Clarridge, J. E., & Sjostedt, A., (2000). Evaluation of PCR-based methods for discrimination of *Francisella* species and subspecies and development of a specific PCR that distinguishes the two major subspecies of *Francisella tularensis. J. Clin. Microbiol., 38*, 4180–4185.

10. Karlsen, F., Kalantari, M., Jenkins, A., Pettersen, E., Kristensen, G., Holm, R., Johansson, B., & Hagmar, B., (1996). Use of multiple PCR primer sets for optimal detection of human Papillomavirus. *J. Clin. Microbiol., 34*, 2095–2100.

11. Klotz, M., Opper, S., Heeg, K., & Zimmermann, S., (2003). Detection of *Staphylococcus aureus* enterotoxins A to D by real-time fluorescence PCR assay. *J. Clin. Microbiol., 41*, 4683–4687.

12. Kowalchuk, G. A., Gerards, S., & Woldendorp, J. W., (1997). Detection and characterization of fungal infections of *Ammophila arenaria* (Marran grass) roots by denaturing gradient gel electrophoresis of specifically amplified 18S rDNA. *Appl. Environ. Microbiol., 63*, 3858–3865.

13. Louie, M., Louie, L., & Simor, A. E., (2000). The role of ADN amplification technology in the diagnosis of infectious diseases. *Can. Med. Assoc. J., 163*, 301–309.

14. Olive, M., & Bean, P., (1999). Principles and applications of methods for DNA-based typing of microbial organisms. *J. Clin. Microbiol., 37*, 1661–1669.

15. Padilla, G. V., (2003). Caracterización molecular de cepas fúngicas productoras de tanasa aisladas en la región semidesértica de Coahuila. BSc dissertation. Universidad Autónoma de Coahuila, Saltillo Coahuila, México.

16. Pérez-Roth, E., Claverie-Martin, F., Villar, J., & Mendez-Alverez, S., (2001). Multiplex PCR for simultaneous identification of *Staphylococcus aureus* and detection of methicillin and mupirocin resistance. *J. Clin. Microbiol., 39*, 4037–4041.

17. Ravenschlag, K., Sahm, K., & Amann, R., (2001). Quantitative molecular analysis of the microbial community in marine artic sediments (Svalbard). *Appl. Environ. Microbiol., 67*, 387–395.

18. Rapley, R., (2000). Recombinant DNA technology. In: Walker, J. M., & Rapley, R., (eds.), *Molecular Biology and Biotechnology* (p. 563). Royal Society of Chemistry: United Kingdom.

19. Reischl, U., (1996). Applications of molecular biology-based methods to the diagnosis of infectious diseases. *Front. Biosci., 1*, 72–77.

20. Relman, D. A., (1998). Detection and identification of previously unrecognized microbial pathogens. *Emerg. Infect. Dis., 4*, 382–389.

21. Restrepo, S., Duque, M., Tohme, J., & Verdier, V., (1999). AFLP`s fingerprinting. An efficient technique for detecting genetic variation of *Xanthomonas axonopodis* pv. manihotis. *J. Gen. Microbiol., 145*, 107–114.

22. Rocha, R. J. C., (2003). Detección de *Pantoea stewartii* directamente de la semilla de maíz usando la técnica inmuno-PCR. BSc dissertation. Universidad Autónoma de Coahuila, Saltillo, Coahuila, México.

23. Rodrigues, J. L. M., Silva-Stenico, M. E., Gomes, J. E., Lopes, J. R. S., & Tsai, S. M., (2003). Detection DNA diversity assessment of *Xylella fastidiosa* in field-collected

plant DNA insect samples by using rRNA and gryB sequences. *Appl. Environ. Microbiol., 69*, 4249–4255.

24. Vinayagamoorthy, T., Hodkinson, R., & Mulatz, K. (2003). Identification of food pathogens using multiple nucleotide based signature sequences. In: *12*[th] *World Congress of Food Science and Technology* (pp. 55, 56). Abstracts of papers. Chicago Illinois USA.

CHAPTER 9

New Molecular Methods for the Detection of Microorganisms

RAÚL RODRÍGUEZ-HERRERA and CRISTÓBAL NOÉ AGUILAR

Bioprocesses and Bioproducts Research Group, Food Research Department, Universidad Autonoma de Coahuila, Saltillo, Mexico, E-mail: raul.rodriguez@uadec.edu.mx (R. Rodríguez-Herrera)

ABSTRACT

The new molecular methods for the detection of pathogens in the food area have had a great growth, these methods are faster, cheaper, and more efficient. This chapter talks about the great impact that molecular techniques have for the detection of pathogens such as the PCR test and some of its variants. In addition, the importance of the use of peptide nucleic acids is mentioned. Finally, we talk about surface plasmon resonance, which is based on biosensors capable of performing a biospecific interaction analysis. These molecular techniques will facilitate the identification of pathogenic microorganisms that are a real problem.

9.1 INTRODUCTION

In recent years, microorganisms have generated great interest in safety, increasing appearance, and spread of virulent bacterial lines and resistance mechanisms against all available antibiotics [1].

The introduction of molecular biology methods in microbiology laboratories helps to obtain sensitive and specific diagnoses in the shortest possible time. These methods are not a substitute for, but a complement to traditional microbiological methods. An integrated analysis is leading to more reliable and effective results. Among all the molecular techniques used, the polymerase chain reaction (PCR) has a great diagnostic value, allowing the detection of etiological agents, and their virulence and resistance genotypes, with great sensitivity and speed [15].

9.2   POLYMERASE CHAIN REACTION (PCR)

This technique is an enzymatic method that allows to copy a specific area of a genome, being able to obtain up to 100,000 copies of it in a test tube. PCR makes use of the enzyme DNA polymerase that is capable of copying DNA molecules. The technique requires that the nucleotide sequence of a region of the desired gene is known, since for PCR to work, it is necessary to have oligonucleotides or short primers (very small pieces of DNA), complementary to the sequences present in the gene or genes of interest from which the DNA polymerase will incorporate complementary nucleotides to the chain that it copies. Similarly, it is necessary that nucleotides in equimolar quantities are present in the test tube (http://www.bio.davidson. edu). This copying process is possible thanks to the use of equipment called thermocycles, which allows the programming of successive cycles with controlled time/temperature gradients. The rationale for PCR can be summarized in the following stages:

- Denaturation of DNA;
- Annealing of initiators;
- Polymerization of complementary chains.

PCR-based methods are frequently used to amplify DNA fragments that are separated by conventional electrophoresis [5]. This technique can sometimes give ambiguous interpretations due to the low specificity of the initiators or contaminations, which produce false positives. Techniques such as nested PCR, real-time (quantitative) PCR, and biosensor technology are efficient in increasing the specificity of microorganism analyzes [10].

9.2.1   MULTIPLE PCR

Many of the PCR products for microorganism detection involve reactions that amplify a single target sequence. Multiple PCR is a variation of the conventional technique where two or more target sequences are simultaneously amplified in the same reaction. This method has great reliability, flexibility, and cost reduction [13].

In the case of multiple PCR, the objective is to simultaneously amplify different specific sequences in a single tube, which necessarily implies that the mixed reagents and the program used are sufficient and adequate to allow the detection of each segment and not inhibit others. To this end, some parameters, such as the concentration of magnesium and initiators, and the type and amount of DNA polymerase, can be adjusted experimentally. For others, a previous exhaustive design must be carried out [15].

In the case of the initiators and the temperature program used, it is necessary to consider several premises: a) choose or design oligonucleotides that do not interact with each other, that is, that do not form oligomers; b) having similar annealing temperatures; c) that each pair amplify a single target sequence, and d) that they generate amplified products of different sizes, which allow them to be separated and differentiated after amplification [15]. An ideal multiple system is one that is capable of detecting target sequences, capable of identifying specific lines and simplify the process of detection and identification of microorganisms [13].

Recent studies have described the use of multiple PCR as a rapid and convenient assay for the detection of microorganisms. However, some modifications have been made to this methodology in order to obtain a more specific reaction. In 2004, James et al. [13] designed a multiple PCR protocol for the detection of the nine transgenic cultures and combinations of primers were used that allowed the identification of specific lines. In one of the identification systems, simultaneous amplification profiling, instead of specific target detection, was used for the identification of four genetically modified (GM) maize lines. Non-specific amplification was used as a reliable tool for the identification of a GM maize line. The cry1A 4–3' primer (antisense) recognizes two sites in the template DNA extracted from transgenic maize event 176 resulting in the amplification of products of 152 bp (expected) and 485 bp (not expected). The last fragment was sequenced and confirmed to be the *cry1A (b)* gene. Simultaneous amplification can identify specific GM lines, as well as the identification of new GM lines containing similar transgenes.

9.2.2   REAL-TIME PCR (RT-PCR)

The use of real-time PCR (RT-PCR) is a practical and fast system as a quantitative method for the detection of microorganisms [2, 12]. This PCR technique has made it possible to quantify the initial amount of nucleic acids during the PCR reaction without the need for subsequent analysis [14]. Furthermore, with fluorescence detection, the amount of DNA synthesized at each moment can be measured during amplification, since the fluorescence emission produced in the reaction is proportional to the amount of DNA formed. This allows the kinetics of the amplification reaction to be known and recorded. The fluorescence detection systems used in RT-PCR can be of two types: intercalating agents and specific fluorochrome-labeled probes [1].

Intercalating agents are fluorochromes that markedly increase fluorescence emission when bound to double-stranded DNA. The most widely used in RT-PCR is SYBR Green I. Specific probes labeled with fluorochromes are probes labeled with two types of fluorochromes, a donor and an acceptor. The process is based on fluorescent resonance energy transfer (FRET) between the two molecules. The most widely used are hydrolysis probes, also called TaqMan probes, molecular beacon probes and FRET probes [1]. Using this method, a target gene can be quantified by preparing a standard curve for known amounts of the gene and by extrapolating the regression line [2].

The combined use of the different PCR methods allows a more efficient analysis, a clear example is the work of Rudi et al. [16] where they designed a multiple PCR based assay for DNA quantification. This method is based on two stages of PCR, in the first cycles the bipartite initiators contained in the 5' termination, a universal sequence and a 3' region specific sequence for each of the GM events are analyzed. Unused initiators are then demoted with a specific exonuclease for single-stranded DNA. The second stage of PCR is performed containing only complementary primers to the universal sequence of the 5' region. Initiator removal is essential for quantitative PCR (qPCR). The oligonucleotides hybridize to internal fragments of the PCR products are then labeled for specific sequences. Hybridization of the labeled oligonucleotides to their complementary sequences in a DNA array allows multiple detection. Where quantitative information is obtained in a range of 0.1–2% for the different GM analyzed. In this work 17 different food and seed samples were examined using a 20-plex system for the simultaneous detection of seven different GM maize lines (Bt176, Bt11, Mon810, T25, GA21, CBH351 and DBT418).

To increase the sensitivity of RT-PCR, the methodology has been modified, for example, in 2003, Zhang et al. [17] designed a new method of RT-PCR using a probe attached to a universal quench. In the design, the universal quenching sequence is approximately 20 bp in size and can hybridize the probe for this quenching, the quenching is attached to the 5' region of a common primer that is specific to the sequence to be amplified. The universal tempering probe is marked with a fluorescent reporter at its 5' termination and with a quencher at its 3' end. During the coupling phase the universal probe joins the 5' end of the universal PCR quenching-initiator and the 3' termination joins the sequence which will be extended. Due to the 5' exonuclease activity of DNA polymerase, the hybridized universal probe is hydrolyzed, promoting the separation of the reporter and the quencher and the generation of a fluorescent signal. In this case the same group of universal sequences and universal probes can be used for different primers.

9.2.3   USE OF CAPILLARY ELECTROPHORESIS (CE) WITH LASER-INDUCED FLUORESCENCE DETECTION

Many of the proposed analytical methods for identifying microorganisms are based on PCR because of the sensitivity, specificity, and applicability of the analysis. The standard procedure consists of separating the PCR products by electrophoresis gels, stained with ethidium bromide, recording the image using an image digitizer, and quantifying the DNA fragments with specialized software. However, these procedures are not very repro-ducible and precise, they require large amounts of sample and are time consuming. In summary, the limited resolution of the electrophoresis gels imposes restrictions on the design of internal standards [5].

The joint use of PCR and capillary electrophoresis (CE) gels seems to be a good alternative for the identification of microorganisms based on DNA analysis. In combination with competitive PCR, CE analysis enables precise detection and amplification of various sequences, as an alternative to RT-PCR. However, UV detection in CE has poor sensitivity and generally cannot be applied to samples with concentrations below 10^{-6} M. The use of induced laser fluorescence in CE dramatically increases the detection limit and dynamic linear range compared to that obtained with UV [6]. This technique involves a high level of automation, uses minimal amounts of samples and reagents, is capable of producing PCR product separations with great efficiency, and it has been proven that it is a good alternative to obtain precise and sensitive results in the quantification of the DNA fragments amplified by PCR [5].

In 2002, García-Cañas et al. [5] using CE gels, detected the presence of transgenic corn in flour. The method is based on the extraction and amplification by PCR of specific fragments of transgenic maize and the subsequent analysis by CE gels with UV detection and induced laser fluorescence. A comparison was made between the two protocols for detecting DNA based on ultraviolet (UV) absorption and by induced laser fluorescence. Obtaining a more sensitive result with the induced laser detector, in a sample of corn flour, contamination was shown that was not detected by UV, which would generate a false positive.

In 2002, García-Cañas et al. [5] compared four different fluorescent intercalating agents for the detection of transgenic corn in flours by CE gels with induced laser fluorescence. The fluorescent intercalating agents compared were: YOPRP-1, SYBR-GREEN-1, ethidium bromide and EnhanCE. SYBR-GREEN-I and YOPRO-1 were shown to give better detection limits than EnhanCE and ethidium bromide. The separation with YOPRO-1 was faster than using SYBR-green-I. YOPRO-1 separations were also more efficient than SYBR-green-1. The cost per analysis and the durability of the buffer is worse with SYBR-green-I than with YOPRO-1. For this reason, YOPRO-1 was selected as the fluorescent intercalating agent, where up to 0.01% of transgenic corn was detected in the flour.

To reduce analysis costs, methodologies have been combined, García-Cañas et al. in 2004 [7], designed a method for the simultaneous detection of five transgenic maize using multiple PCR, followed by CE with laser-induced fluorescence. In this method, a hexaplex PCR protocol was used for the amplification of five varieties of transgenic corn (Bt11, T25, MON810, GA21, and Bt176), and the use of electrophoresis gels from laser-induced fluorescence was shown to be very useful and informative for the optimization of multiple PCR parameters such as extension time, concentration of the PCR buffer and primers. This method is very sensitive and also solves false positive problems.

9.2.4 MICROARRAYS

Microarray technology can increase the analysis speed of PCR products. DNA microarrays are an analytical system that allows the simultaneous detection of many nucleic acid sequences in a sample. Each DNA sequence is represented by a covalent bond from an oligonucleotide probe on the modified surface of a piece of glass. The probes in the array are hybridized with the fluorescent marker for PCR products. The analysis of the scanner reveals the presence of the labeled material containing the complementary sequences of those marked in the microarray [10].

9.3 PEPTIDE NUCLEIC ACIDS (PNAS)

Peptide nucleic acids (PNAs) are analogs of oligonucleotides where sugar-phosphate has been replaced by the pseudo peptic chain of the N-aminoethylglycine monomer [10]. Unlike nucleic acids, PNAs do not contain pentose or phosphate groups. The main advantage of these biometric molecules compared to their natural analogs is their great affinity to establish links with DNA chains. The lack of electrostatic repulsion between them makes these bonds stronger than those between two strands of DNA [11].

In 2004, Germini et al. [9] reported the combination of microarray technology with PNAs for the detection of GM soybeans. Different PNAs were designed, synthesized, and covalently linked to the functional surfaces to construct a microarray and to identify the constitutive gene for lectin and the region of the gene corresponding to the Roundup Ready GM soybean. The effect of the PNA length on signal intensity and specificity was evaluated as well as the detection conditions of both single and double chain PCR products. The best results were obtained with the single chain PCR products and with long PNAs. The advantages of this method over other technologies are: 1) PNAs are more efficient and form more stable hybrids than oligonucleotides, 2) PNAs are a highly specific sequence, and 3) microarrays allow simultaneous molecular analysis of many sequences.

9.4 SURFACE PLASMON RESONANCE (SPR)

This method is based on biosensors capable of performing a biospecific interaction analysis (BIA) for monitoring a variety of molecular reactions in real time. It is an optical technique that detects and quantifies changes in the refractive index near a surface of the sensor chip in which the ligands are immobilized, allowing the detection of biomolecules (analytes) inter-acting with the ligand. If the ligand is single-stranded biotin-labeled DNA, SPR technology can easily monitor DNA-DNA hybridization in real time [4].

Surface plasmon resonance is an optical phenomenon that occurs when polarized light is directed from a layer with a higher index of refraction (a prism) towards a layer with a lower index of refraction, which in this case is a metallic layer of gold or silver, which is located between the prism and the sample. The light that hits the interface between the metal and the prism causes the excitation of a surface plasmid for a certain angle of incidence of light, which strongly depends on the refractive index of the medium adjacent to the metal sheet, so the variations that occur in it will be detected as changes in the angle that will be proportional to the concentration. It can also be used with labeled fluorescent molecules. This has the advantage of an improvement in relative sensitivity due to internal total fluorescence reflection. The instrumentation that requires this type of analysis is a sensor, an SPR detector, software for the control and treat-ment of data, and a system to introduce the reagents and analytes on the sensor surface [11].

In 2002, Feriotto et al. [3] demonstrated that the BIA using surface plasmon resonance and biosensor technologies is an easy, fast, and automatic proposal for the detection of genetically modified organisms (GMOs). The first proposal (Figure 9.1(A)) is based on the immobilization of the single-chain synthetic oligonucleotide target in the sensor chip and the injection of the specific DNA sample different in length. This format is of great importance for the study of relationships between probe lengths, the efficiency of the hybridization of the target DNA and the stability of the generated molecular hybrids. The second, third, and fourth formats are relevant for diagnostic purposes and are based on PCR by amplifying the sequences of the genes that want to be identified. In Figure 9.1(B), PCR is performed using a biotin-labeled primer immobilized on a sensor chip and analyzed by injection of a compatible oligonucleotide. In Figure 9.1(C) the probes are immobilized on the sensor chip and the asymmetric PCR product to be analyzed is injected. In Figure 9.1(D) the samples generated by PCR are immobilized on the sensor chip and the asymmetric products to be analyzed are injected. In this technology, the fact that radioactive markers are not needed is very important, the procedure is carried out in real time, and small amounts of the ligand and analyte are required to obtain results.

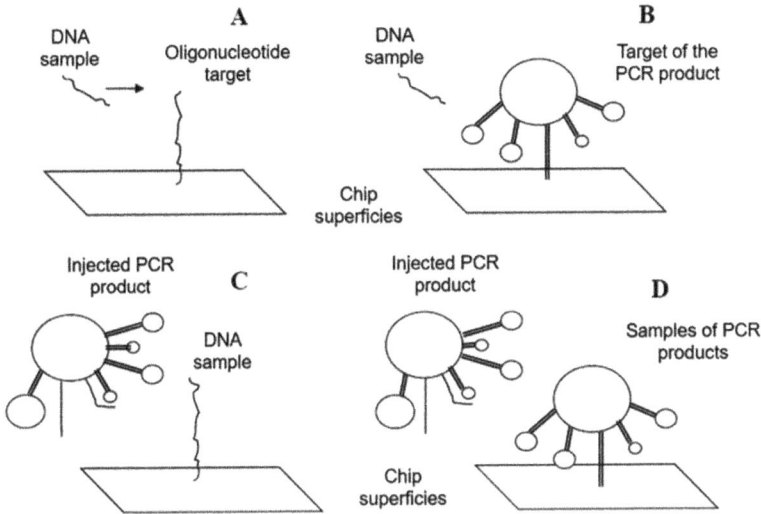

FIGURE 9.1 Experimental strategies and formats for the detection of GMOs using BIA-based SPR and chip sensors that transport oligonucleotides or PCR products from lectin or roundup ready.

For 2003, Feriotto et al. [4] designed and performed an SPR protocol based on a BIA not only quantitative but also multiple determination of GMOs. The protocol is based on the immobilization of the multiple PCR products in a simple streptavidin-coated cell flow sensor chip and the injection of specific probes. Zein biotin-labeled and Bt-176 products were immobilized on the SA sensor chip taking advantage of the streptavidin-biotin interaction. The approach is based on the immobilization of the same cell flow of two targets of PCR products obtained by multiple PCR using biotin-labeled primers and the following analysis by consecutive injection of a compatible nucleotide probe. To produce the double stranded sequences of the target gene, multiple PCR had to be carried out using standard genomic DNA with an excess of Bt-R and ZM-r primers with their respective biot-Bt-F and biot-ZM-F marked with biotin. This was done to minimize the presence of biotin labeling of unincorporated initiators in the PCR mixture. The zein and Bt-176 multiple PCR end products were purified with Microcon-30. Agarose gel analysis and direct sequencing of the PCR products confirmed the specificity of the PCR reaction. Results were compared with Southern blot and qPCR using an ABI Prism 7700.

KEYWORDS

- **biospecific interaction analysis**
- **capillary electrophoresis**
- **fluorescent resonance energy transfer**
- **genetically modified organisms**
- **peptide nucleic acids**
- **polymerase chain reaction**

REFERENCES

1. Costa, J., (2004). Reacción en cadena de la polimerasa (PCR) en tiempo real. *Enferm. Infecc. Microbiol. Clin., 22*, 299–305.
2. Ding, J., Jia, J., Yang, L., Wen, H., Zhang, C., Liu, W., & Zhang, D., (2004). Validation of a rice specific gene, sucrose phosphate synthase, used as the endogenous reference gene for qualitative and real-time quantitative PCR detection for transgenes. *J. Agric. Food Chem., 52*, 3372–3377.
3. Feriotto, G., Borgatti, M., Mischiati, C., Bianchi, N., & Gambari, R., (2002). Biosensor technology and surface plasmon resonance for real-time detection of genetically modified roundup ready soybean gene sequences. *J. Agric. Food Chem., 50*, 955–962.
4. Feriotto, G., Gardenghi, S., Bianchi, N., & Gambari, R., (2003). Quantitation of Bt-176 maize genomic sequences by surface plasmon resonance- based biospecific interaction analysis of multiplex polymerase chain reaction (PCR). *J. Agric. Food Chem., 51*, 4640–4646.
5. García-Cañas, V., González, R., & Cifuentes, A., (2002a). Detection of genetically modified maite by polymerase chain reaction and capillary gel electrophoresis with UV detection and laser induced fluorescence. *J. Agric. Food Chem., 50*, 1016–1021.
6. García-Cañas, V., González, R., & Cifuentes, A., (2002b). Ultrasensitive detection of genetically modified maize DNA by capillary gel electrophoresis with laser-induced fluorescence using different fluorescent intercalating dyes. *J. Agric. Food Chem., 50*, 4497–4502.
7. García-Cañas, V., Cifuentes, A., & González, R., (2004a). Quantitation of transgenic Bt event-176 maize using double quantitative competitive polymerase chain reaction

and capillary gel electrophoresis laser-induced fluorescence. *Anal. Chem., 76,* 2306–2313.

8. García-Canas, V., González, R., & Cifuentes, A., (2004b). Sensitive and simultaneous analysis of five transgenic maizes using multiplex polymerase chain reaction, capillary gel electrophoresis and laser- induced fluorescence. *Electrophoresis, 25,* 2219–2226.

9. Germini, A., Zanetti, A., Salati, C., Rossi, S., Forre, C., Schmid, S., & Marchelli, R., (2004a). Development of a seven-target multiplex PCR for the simultaneous detection of transgenic soybean and maize in feeds and foods. *J. Agr. Food Chem., 52,* 3275–3280.

10. Germini, A., Mezzelani, A., Lesignoli, F., Corradini, R., Marchelli, R., Bordoni, R., Consolandi, C., & Bellis, G., (2004b). Detection of genetically modified soybean using peptide nucleic acids (PNAs) and microarray technology. *J. Agr. Food Chem., 52,* 4535–4540.

11. González, V., García, E., Ruiz, O., & Gago, L., (2005). Aplicaciones de Biosensores en la Industria Agroalimentaria; Informe de Vigilancia Tecnológica. Madrid, España.

12. Hernandez, M., Duplan, M., Berthier, G., Vaitilingom, M., Hauser, W., Freyer, R., Pla, M., & Bertheau, Y., (2004). Development and comparison of four real-time polymerase chain reaction systems for specific detection and quantification of *Zea mays* L. *J. Agric. Food Chem., 52,* 4632–4637.

13. James, D., Scmidt, A., Wall, E., Green, M., & Masri, A., (2004). Reliable detection and identification of genetically modified maize, soybean and canola by multiplex PCR analysis. *J. Agric. Food Chem., 51,* 5829–5834.

14. Mason, G., Provero, P., Veira, A. M., & Accotto, G. P., (2002). Estimating the number or integrations in transformed plants by quantitative real-time PCR. *BMC Biotechnol., 2,* 1–10.

15. Méndez-Álvarez, S., & Pérez-Roth, E., (2004). La PCR Múltiple en Microbiología Clínica. *Enferm. Infecc. Microbiol. Clin., 22,* 183–192.

16. Rudi, K., Rud, I., & Holck, A. A., (2003). Novel multiplex quantitative DNA array based PCR (MQDA-PCR) for quantification of transgenic maize in food and feed. *Nucleic Acids Res., 31,* e62.

17. Zhang, Y., Zhang, D., Li, W., Chen, J., Peng, Y., & Cao, W., (2003). A novel real-time quantitative PCR method using attached universal template probe. *Nucleic Acids Res., 31,* 20.

18. *Integrating Concepts in Biology.* (2021). http://www.bio.davidson.edu/113 (accessed on 28 September 2021).

CHAPTER 10

Fungal Production and Function of Phytase

ALBERTO A. NEIRA-VIELMA,[1] CRISTÓBAL NOÉ AGUILAR,[2]
ANNA ILYINA,[2] GEORGINA MICHELENA-ÁLVAREZ,[3] and
JOSÉ L. MARTÍNEZ-HERNÁNDEZ[1]

[1]*Nanobioscience Group, School of Chemistry, Autonomous
University of Coahuila, C.P. – 25280, Saltillo, Coahuila, México,
E-mail: aneiravielma@uadec.edu.mx (A. A. Neira-Vielma)*

[2]*Bioprocesses and Bioproducts Research Group, Food Research
Department, School of Chemistry, Autonomous University of Coahuila,
C.P. – 25280, Saltillo, Coahuila, México*

[3]*Cuban Institute for Research on Sugarcane Derivates,
Postal Zone 10, San Miguel del Padrón, La Habana City, Cuba*

ABSTRACT

The growing food demand due to the increase and demand of society it has given a fundamental approach to the composition of animal diets, which must be prepared according to the requirements that the animal needs and its purpose (high production of sub-products or weight gain for meat production). Therefore, the study of the conditioning of raw materials with vegetable origin and the production of additives to promote and improve the use of allowances sold in the market has increased. In vegetable product, mainly cereals and grains, which are the raw materials for these diets, are contained in various proportions; factors involved in the use of nutrients in the gastrointestinal tract, mainly in non-ruminants. One, perhaps the most abundant is the phytic acid which forms complexes of high molecular weight with various nutrients and inhibits some digestive enzymes (mostly amylases cells) causing malabsorption of nutrients and a deficit in digestion. In order to improve the hydrolysis of this compound and the use of nutrients, it has been found that the addition of exogenous phytase (myo-inositol hexaphosphate phosphohydrolases) mainly from fungal organisms represents a significant efficacy in animal feed. These enzymes belong to the family of histidine acid phosphatases, heterogeneous group of proteins, which carry out hydrolysis reactions in the phosphoester bonds of phytic acid molecule. Recently, due to increasing tendency biotechnology and environmental impact are sought various bioprocess, which carry the production of said enzyme, which is promising to date. The purpose of writing this, is to show the importance of phytase in improving animal nutrition. The main sources of obtaining, and catalytic function level. To encourage the use of efficient, economic, and sustainable methods. As are the biotechnological methods. For the improvement of some specific sectors.

10.1   INTRODUCTION

The raw material of vegetable origin, mainly legume seeds and cereal grains, has phytic acid, which is one of the main anti-nutritional factors that affect the metabolism of ruminants [1]. This compound is an organic acid comprising a molecule of myoinositol and six molecules of phosphate. It has the ability to form highly stable ionic bonds with minerals causing the formation of insoluble chelates that cannot be absorbed by the organism [2]. This compound is used in various metabolic processes of the plant cell, such as the synthesis of cell wall polysaccharides and other tissues and organs in various plants (pollen, roots, and tubers). It is considered the main source of phosphorus in seeds, serving as storage and retrieval functions of phosphorous during the germination and development of the plant [3–5]. Current diets on the market are mainly prepared from raw materials containing significant amounts of this compound (from 1% to 6%) [6–8]. To try to remedy this problem, methods are sought to reduce the adverse effects of this compound or eliminate it altogether. Therefore, research around biotechnology is a high impact capo in the agribusiness sector. An alternative, perhaps the most efficient so far and that has proven to keep the properties of the food, is the use of enzyme phytase hydrolase type which is approved by the Food and Drug Administration (FDA) United

States of America, for use as a functional additive in feed for poultry and swine [9, 10]. These enzymes act hydrolyzing the phytic acid to inositol phosphate derivatives, which have a smaller number of phosphate and even free inositol, which have a lower ability to bind minerals [11]. This enzyme is naturally present in the digestive tract of certain animals, plants, and produced by microorganisms such as bacteria, yeasts, and filamentous fungi largely of which some are shown in Table 10.1 [12].

TABLE 10.1 Producing Microorganism, Method, and Substrate

Microorganism	Production Technique	Substrate	References
Bacteria			
B. subtilis	SmF	–	[13–15]
Escherichia coli	SmF	Complex medium	[16]
E. coli	–	–	[17]
Lactobacillus amylovorus	SmF	Glucose	[18]
Pseudomonas sp.	–	–	[19]
Prevotella ruminicola	–	–	[20]
Selenomonas ruminantium	–	–	[21]
Yeasts			
Schwanniomyces castellii	SmF	Wheat bran, cotton our	[22]
S. occidentalis	–	–	[23]
Fungi			
Aspergillus sp.	SmF	Complex medium	[24]
A. ficuum	SSF	Canola meal	[25]
A. niger	SmF	Maize starch	[3]
A. niger	SmF/SSF	Complex/wheat bran	[26]
A. oryzae	–	–	[27]

Producing microorganisms include the filamentous fungus of the genus *Aspergillus* because diver's studies confirmed they produce the phytase enzyme more active and top characteristics of both pH and temperature which results in stability of extracellular phytase. Which puts the microorganisms of the genus *Aspergillus* as the most used in industrial production of this enzyme [28–30]. Between phytase producing microorganisms the filamentous fungus genus Aspergillus, which produces the most active and stable enzyme pH and temperature. Therefore, microorganisms of the genus *Aspergillus* the most used in industrial production of this enzyme [30, 31]. Enzymes are among the products of interest obtained from microbial sources. The solid-state fermentation (SSF) in a process potential enzyme interest production [32]. Currently a lot of companies have developed processes to produce phytase, bacterial, and fungal. Table 10.2 shows the main level to world and products they offer.

As described previously many microorganisms including filamentous fungi can produce the phytase. These possess the ability to grow almost any on surface plant residues and almost any type of material organic with some moisture, particularly those with a relatively low water activity (WA) therefore is the method most used today for the production of this metabolite [33, 34].

TABLE 10.2   Main Phytase Production Companies and Their Products

Company	Product
Adisseo France S.A.S	**Rovabio®:** Enzyme preparation that improves the digestibility of certain raw animal level subjects, birds, and pigs.
AB Vista	Quantum® from AB Vista: Phytase enzyme preparation, for the benefit of monogastric animals. It is effective and Presis for controlled liberacipin in the intestinal tract. Operates over wide ranges of pH, including the upper gastrointestinal tract, is stable and thermotolerant.
BASF SE	**Natuphos®:** Contains at least 5,000 units of phytase (FTU) per gram, guaranteed.
Advanced Enzymes Technologies Limited	**SEBFeed Phy 5:** Thermostable phytase granulate. Guaranteed to reduce phosphorus in poultry and swine.
Royal Dsm N.V.	**RONOZYME® Phytases:** Combination of protease and phytase to improve animal digestion.
BIOMIN Holding GmbH	**Biomin® Phytase:** It is the third generation of phytase. It has high efficacy in improving the bioavailability of phosphorus and other nutrients in feed.

10.2   PHOSPHORUS CONTENT IN RAW MATERIALS FOR CATTLE FEED

The organic phosphorus in plant represents the largest fraction forming phytic acid, which is the most abundant phosphoglucose. By contrast in animals' products, the predominantly is inorganic phosphorus as ortho-phosphates [35].

In the classification of phosphorus metabolism, phytic acid is an important factor antinutritional. An organic acid comprising six phosphate molecules, and an inositol [36]. According to studies, the digestibility of inorganic and organic phosphorus is similar and near to 100% (80–100). Digestibility of phytate phosphorus in no-ruminant organisms is null so it is impossible to use [37].

Something very important is the variation of phosphorus content in the raw materials. From a homogeneous view, forage grasses have higher amounts than legumes, and seeds (cereal grains, legumes, and oilseeds) more forages [38].

The level of P varies not only between gender also within species. In raw materials of plant, origin depends on the P content of the soil, cultivar, ripeness, growing conditions and weather, among others [39, 40]. On the other hand, the level of phosphorus in mineral supplements depends on multiple factors such as the source material, manufacturing process and the degree of hydration of the material in question [41].

The amount of P in the raw materials is divided in inorganic mainly orthophosphate highlighting, and organic, forming important molecules as adenosine triphosphate (ATP), nucleic acids, phospholipids, phospho-proteins, and phosphoglucose. Which form the more minimal structure of the raw material a cellular level [35, 42]. Those organisms that degrade fully or partially organic P in tract gastrointestinal, promote PO_4 release. Which is the only way the organism can use the full P [42, 43]. In the vegetal ingredients forming the market diets, the organic P represents the mayoritarira fraction in form of phytic acid which comprises up to 80% of total [38, 44]. It is estimated that 60–80% of the total P content in the grains and byproducts forming the phytic acid and its salts, having greater affinity for metals like Ca^{++}, K^+ and Mg^{++} [35, 38, 43, 44].

Due to this deficiency in the vegetal material, supplement the diets marketed with P. In order to increase the use of P on the organism and its bioavailability. Are used for this orthophosphate, Na^+, Ca^{++}, K^+, NH_4^+ and combinations which provide a small amount of P, however mow still insufficient and in some cases may significantly raise the cost [42, 45, 46].

To ensure a good amount of P in diet and know the amount that the raw material contains in their two fractions is most importantly a precise analytical determination. This is achieved through various methodologies mainly those attached to the normativity of the AOAC [88], detailing methods include detection of both fractions of P [36, 38, 42]. Important information to decide the amount of P to supplement in dietary, because in some cases 70% of P detected is phytic acid [44]. Table 10.3 shows disproportionately more plant materials used to develop animal feed in which the above is appreciated.

TABLE 10.3　Phosphorus Content in Some Raw Materials (g/Kg MS) [47]

Source Materials	Total	Phytate	Inorganic	Nonphytate	Available
Corn	3.0	2.0	1.2	1.0	0.9
Oat	3.6	1.9	1.2	1.7	1.1
Triticale	4.4	3.0	1.5	1.4	1.3
Grain	3.9	2.2	1.1	1.7	1.2
Barley	4.3	2.2	1.1	1.7	1.2

This type of data is important to assign a precise value of available P content. Of great value in diets for non-ruminants, where should be considered the nature of the P present in raw material, both total phosphorus and phytic phosphorus [45].

10.3　PHYTIC ACID AND PHYTATES IN FEED MATERIALS OF PLANT ORIGIN

Phytates (complex formed by binding of inositol hexaphosphate mainly to metals) are the most abundant source of P storage in seeds and plant material [48]. Preferentially accumulate in the seed, its content in stems and leaves is low. The location varies by type of grain, wheat, and rye between 80 and 90% are located in the aleurone and pericarp, maize, and sorghum accumulate in the germ [43, 44]. In legumes has been identified concentrate in cotyledons, and distribution oleaginous in tends to be diffused throughout the seed associated to globular bodies with high protein [42, 49].

This complex identified initially as a source of P during germination; however, other studies confirm that this is extremely important in the seed for the prevention of oxidative stress and preventing the death of the embryo [50]. Its function is directly related to the ability of phytic acid to chelate minerals such as Fe^{+++}, Ca^{++} and Zn^{++}, this reduces the formation of oxidative compounds that may be created during germination [51].

Phytates present in various raw materials have certain distinctive characteristics such as pH, stability, and even the location of the compound in the material. These characteristics influence the solubility and therefore the hydrolysis, which is closely related to the release of P [49, 52]. Important fact to pretreat the raw materials to achieve decrease the amount of this compound [53]. The industrials main treatments to reduce the presence of phytic acid are chemical methods, such as alkaline and physical treatment such as heating and extrusion. However, these types of treatments have shown greatly affect the quality of food and their nutritional value [39, 54]. A new option is the enzymatic treatment of the raw materials and the incorporation therein of an enzyme capable of reducing levels of this compound. These favors increasing the bioavailability of P and other nutrients, without altering the nutritional characteristics [37, 45, 55, 56].

Phytic acid and the complex derived phytate are important as a source of potential nutritionally P [35, 37]. In ruminants a fermentative process originates at the gastrointestinal level, whereby the phytate is degraded by bacteria in the rumen flora, and results in the release of P for subsequent absorption by the animal [37, 57]. For this reason, various studies classified the as source phytate of non-digestible phosphorus for no ruminants' organisms [35, 58].

At level of gastrointestinal tract in ruminants, the P present as part of phytate has low availability. Because the organism consumes, primarily poultry and pigs lack the enzyme necessary to degrade the compound or at least in significant amounts which promote degradation, and the subsequent use of the P [36, 37, 58]. So, a lot of phytic acid appears in the feces of the animal, which is of great prejudice, as decrease the quality of diets and rises problems of environmental pollution [42, 58].

The stability of phytic acid is due to capacity of forming high strength bonds with metals, and even high molecular weight compounds, such reactions originate at the level of intestinal tract [7, 51].

Another feature of phytic acid, can form insoluble complexes with proteins and starch [59]. The liaison formed by this compound are ionic and are regulated by the pH of the medium [58, 59]. At low pH, phytic acid binds with basic residues such as lysine, arginine, or histidine to form highly stable insoluble complexes [37, 60]. The complexes formed change the structure of the protein, reducing its solubility, digestibility, and functionality [40, 61]. These complexes decrease digestibility of nutrients to the which it is attached, also derive an endogenous response by the animal [58, 62]. The combination mechanisms cause in drastic anti-nutritional effects which are directly associated with the interaction capabilities of phytic acid [37, 63]. One of these mechanisms, maybe the more harmful in animal nutrition is the ability of phytic acid to inhibit enzymes crucial in the digestive processes, such as α-amylase, protease, trypsin, pepsin, and lipase, inhibitions by interactions still unknown [64].

The advantage of the study of this compound is that it can get to know more specific location, their mechanisms of action, the role within the organization and how they interact under various conditions [44, 49]. This opens an opportunity to improve animal performance, beyond seeing only the phytate as a phosphorus source [37].

10.4　PHYTASE AS AN OPTION FOR DECREASING PHYTIC ACID CONTENT IN PLANT MATERIALS

Phytases are part of the sub-family of enzymes acid phosphatase, of the hydrolase type. These act by breaking bonds phosphomonoester, forming myoinositol hexaphosphate and inorganic phosphorus compounds. That have less or no effect antinutricional and chelator [3, 65]. Natural sources of phytases are highly varied, some plants (cereals, legumes, tubers, etc.), The intestine of certain non-ruminants, certain of the microflora microorganisms the ruminant animals and other organisms included *Aspergillus*, *Bacillus*, *Pseudomonas*, yeasts, etc. [65–67].

The hydrolysis of phytate by action of phytase, improvement the absorption and retention of P, Ca^{++}, Mg^{++}, Zn^{++}, Cu^{++}, Fe^{+++} and amino acids, especially in deficient diets [47, 59, 68]. This has shown reduced by up to 50% of the pollution problem, if used foods supplemented are phytase. In ruminant dung and partially in animals such as rabbits, the phytase production by microorganisms in the rumen and cecum is abundant so the use of phytic acid is not a problem [64]. However, the problem lies specifically in non-ruminants lacking this enzyme, so the interest in producing industrially phytases is high. Therefore, phytases of microbial origin have received increased attention by the high potential, in the animal feed industry.

This has led to the development of various biotechnological processes for the production of phytase industrially, mainly from fungal strains among which *Aspergillus* species excels, microorganisms known for its extensive enzymatic machinery [29, 30, 59, 69]. The phytases of fungal origin have two main advantages over those of bacterial origin: the optimum pH of phytase activity of fungal origin is between 2.5 and 5.5, values similar to those held by the gastrointestinal tract of non-ruminants, unlike bacterial phytases, having optimum pH of activity near neutral [3, 45, 70]. Besides, the recovery process of the fungal phytase is carried out more easily because they are in their most extracellular unlike the bacterial, which are intracellular, except those of the genus Bacillus.

This makes the purification processes phytases of fungal origin are economically more viable [71, 72]. The microorganisms able to grow on this condition are mainly fungus. Which have proven to be good producers, even in to industrial scale processes as well as being less susceptible to contamination and abrupt physicochemical changes in the system [72–74].

10.5 PHYTASE PRODUCTION FOR USE AS A FOOD ADDITIVE

Since phytic acid is one of the main sources of phosphorus storage and some amounts of carbon, is a potential reserve for plants in which is located [8, 44, 52]. To promote the dephosphorylation of this compound, increase the bioavailability of the nutrients that are attached to it and take use of its basic structural composition, as mentioned previously the most viable is the use of phytase enzyme [32, 74, 75].

Generally, the main advantage of the phytases produced from fungi is that they are extracellular, unlike those produced by bacteria, which are intracellular except for the ones produced by *Bacillus subtilis* and *Enterobacter* [72, 76]. However, the production of phytase depends mainly on the characteristics that want to be promoted in the enzyme, based on the source of purity and foremost you want to produce [27, 77].

10.5.1 PLANT PHYTASES

In the case of the production of phytase in plants have been identified and studied mainly in rice, wheat, corn, soy, and rye [43, 78]. Studies done on these materials have shown that both phytic acid and phytase are key in germination; because of the degradation of this compound the plant receives an extra contribution of phosphorus. During development of the plant, the phytase activity is increased up to 10 times more [7, 49].

10.5.2   BACTERIAL PHYTASES

Moreover, in production of phytases bacterial, have identified various microorganisms responsible, mainly *B. subtilis*, *E. coli*, *Pseudomonas*, and *Klebsiella* [27, 57, 79]. In this type of microorganism, expression of this protein has the function of phytate degradation, as auxiliary source of nutrients [27, 89]. These strains usually tend to be isolated in soil near high in organic matter.

10.5.3   FUNGAL PHYTASES

Due to the ability of fungi to metabolize a large number of raw materials. Has greatly increased interest in the isolation and identification of novel strains producing this enzyme, mainly including species of *Aspergillus*, *Mucor*, *Penicillium*, and *Rhizopus*, microorganisms that have been tested for production showing good result [27, 32, 63]. A unique feature, is that all producing fungus of the enzyme, expressed a type of phytase extracellular highly active and stable [3, 30]. Of these microorganisms, *A. niger* was identified as the active most producer of fungal phytase, being overcome with minimal difference by *A. ficuum*. Both produce an enzyme exhibiting the highest rate of activity at pH 2.5, 5.0, and 6.0, and an optimum temperature of 63°C, the latter being susceptible to change depending on the microorganism which produces it [56, 90].

10.5.4   BIOTECHNOLOGY PHYTASES (PRODUCTION AND USE)

Over time, the use of fungi for the production filamentous enzyme has increased rapidly and phytases are no exception. The industrial production level of this enzyme has been achieved by use of fungal microorganisms, fermentation cultures mainly in solid state and less frequently in liquid fermentation [29, 50, 74]. The solid fermentation is a process in which the low moisture content (40 to 80%) plays an important role, at the cellular level as to level physicochemical [28]. To develop such processes have been analyzed mainly vegetable substrates, which meet minimum nutritional characteristics and preferably contain high amounts of phytate, which promotes the synthesis of the enzyme of interest. The substrates that have shown to these features are wheat bran, soybean meal, corn meal and canola meal [35, 55, 75]. In published studies using the raw materials listed, in solid fermentation processes, the production of this enzyme shown to be highly favorable. Due to the features provided by these substrates such as support and encouragement of the physicochemical processes, transfer of heat, oxygen, and nutrients [26, 29, 76]. An advantage of fungal micro-organisms used for the production of phytases, filamentous metabolism is that this is not affected by the amounts of anions formed in the middle, as in the case of yeast as *Schwanniomyces castellii*, microorganism phytase producing highly active susceptible to be affected by the above [65, 80]. Therefore, special interest in producing this type of enzymes, ensuring always maintain its stability and activity until time of use is provided.

Due to favorable characteristics that phytase possessed and its practical applications as an additive in diets of no ruminant this enzyme has taken a position of great interest for biotechnological applications for the reduction of phytate content in fodder and commercial foods [77]. Phytases used as an additive must prove that they are effective in releasing phosphate from phytic acid and must demonstrate their effectiveness at the level of digestive tract both to withstand the conditions of pH and temperature, to perform this action [7, 81, 82].

As the enzyme is used in this process must show their strength characteristics of heat inactivation of both food processing and of storage, finally something that is very important is the cost benefit for the use of this additive, which must be economically feasible for both the producer and the consumer [28, 37, 68, 72].

From the features above mentioned the thermo stability is essential, this due to use post-production which is precisely the field application, since its main use is as granules, this method is often performed at temperatures between 65 and 95°C [83, 84]. Therefore, the production of this enzyme is not the main challenge, this occurs in both the recovery and the formulation for which is mainly used as the chemical to coating phytase, process that favors the thermal stability of the enzyme, being candidates for food supplements [40, 65]. Therefore, it has been discarded that phytases from other sources such as plant are useful in these processes, due to be inactivated susceptibility to the slightest changes in temperature and pH [71].

Because of this and according to demands of the market to favor the demands of feed search of microorganisms and substrates suitable for the production of phytase is an important point in biotechnological applications aimed at agribusiness [37].

Besides this the current biotechnological process with the support from other sciences such as genetic engineering and proteomics seek the isolation, for production or otherwise making the perfect biotech phytase, making it have thermal stability, it has tolerance to storage, transport long term, and stability under environmental variation [12, 29, 45, 56].

The submerged fermentation (SMF) has been used as the main potion in production technology. Recently SSF has positioned itself to the production of primary and secondary metabolites [31, 32, 69].

Several authors have discussed the potential SSF, highlighting the benefits brought for a lot of areas of work. Examples include bioremediation and biodegradation of hazardous compounds, detoxification of chemical residues, biotransformation, and nutritional enhancement of crops and the production of biologically active secondary metabolites, including antibiotics, alkaloids, growth of the plant enzymes, organic acids, bio-pesticides, etc. [25, 85, 86]. Until a few years ago the SSF, were a subject of limited investigation as it considered as "low-tech systems." Which in recent years was discarded because it seems to be promising for the production of value added, generating products 'low cost' high volume [12, 56].

The vast production of agro-industrial waste due to the increase of the world population, is a latent advantage in this type of process [32, 87].

And there is no exception to produce phytase, by the culture conditions, the nature of the substrate and availability of nutrients in sayings of substrate are favorable for production factors [2, 87].

10.6   CONCLUSION

Therefore, been given importance in the use SSF with filamentous fungi for the production of commercially important metabolites. Moreover, its benefits lie in its simplicity in terms of equipment and control less expensive process, adding that often results in higher yields compared to SMF.

The wide application panorama and overall benefit of the biotechnological enzyme derived from environmental, farming, and crop production areas sets the ground floor for the evolution and improvement of the before mentioned areas in the use of this enzyme. Moreover, without dough something that needs to be considered in the use of this enzyme is that without the proper study of the production cycle to ensure an accessible price to all processes that imply the use of phytases base products.

The main features to win, coupled with the cost, is the thermal stability and the very stability of the enzyme and/or enzyme preparation used as a food additive. As this industry is the major stakeholder for the particular use of the enzyme; Therefore, to improve the bioavailability of minerals and nutrients by reducing the phytate content of a given food, and secondly to produce functional food.

All this provides a point each day of study for biotechnological processes and boost them, which is of significant importance for the improvement of the livestock field coupled with the added value given to agribusiness residues that are the main supports used for the production of this enzyme.

KEYWORDS

- **adenosine triphosphate**
- **bioprocess**
- **fermentation**
- **phytase**
- **phytic acid**
- **solid state fermentation**

REFERENCES

1. Menezes-Blackburn, D., Gabler, S., & Greiner, R., (2015). Performance of seven commercial phytases in an *in vitro sim*ulation of poultry digestive tract. *J. Agric. Food Chem., 63*, 6142–6149.
2. Bhavsar, K., Kumar, V. R., & Khire, J. M., (2010). High level phytase production by *Aspergillus niger* NCIM 563 in solid state culture: Response surface optimization,

up-scaling, and its partial characterization. *J. Ind. Microbiol. Biotechnol., 38,* 1407–1417.

3. Kim, T., Mullaney, E. J., Porres, J. M., Roneker, K. R., Crowe, S., Rice, S., Ko, T., et al., (2006). Shifting the pH profile of *Aspergillus niger* PhyA phytase to match the stomach pH enhances its effectiveness as an animal feed additive. *Appl. Environ. Microbiol., 72,* 4397–4403.

4. Ma, X. F., Tudor, S., Butler, T., Ge, Y., Xi, Y., Bouton, J., Harrison, M., & Wang, Z. Y., (2011). Transgenic expression of phytase and acid phosphatase genes in alfalfa (*Medicago sativa*) leads to improved phosphate uptake in natural soils. *Mol. Breed., 30,* 377–391.

5. Raboy, V., (2003). myo-Inositol-1,2,3,4,5,6-hexakisphosphate. *Phytochem., 64,* 1033–1043.

6. Graminha, E. B. N., Gonçalves, A. Z. L., Pirota, R. D. P. B., Balsalobre, M. A. A., Da Silva, R., & Gomes, E., (2008). Enzyme production by solid-state fermentation: Application to animal nutrition. *Anim. Feed Sci. Technol., 144,* 1–22.

7. Jondreville, C., Schlegel, P., Hillion, S., Chagneau, A. M. M., & Nys, Y., (2007). Effects of additional zinc and phytase on zinc availability in piglets and chicks fed diets containing different amounts of phytates. *Livest. Sci., 109,* 60–62.

8. Wang, H., Zhou, Y. Y., Ma, J., Zhou, Y. Y., & Jiang, H., (2013). The effects of phytic acid on the Maillard reaction and the formation of acrylamide. *Food Chem., 141,* 18–22.

9. Rodrigues Da, L. J. M., Albino, P. S., Pereira, T. D. P., Dias, N. M., Soares Da, S. J., Cuquetto, M. H., & Megumi, K. M. C., (2013). Production of edible mushroom and degradation of antinutritional factors in jatropha biodiesel residues. *LWT - Food Sci. Technol., 50,* 575–580.

10. Rutherfurd, S. M., Chung, T. K., & Moughan, P. J., (2014). The effect of dietary microbial phytase on mineral digestibility determined throughout the gastrointestinal tract of the growing pig fed a low-P, low-Ca corn-soybean meal diet. *Anim. Feed Sci. Technol., 189,* 130–133.

11. García-Mantrana, I., Yebra, M. J., Haros, M., & Monedero, V., (2016). Expression of bifidobacterial phytases in *Lactobacillus casei* and their application in a food model of whole-grain sourdough bread. *Int. J. Food Microbiol., 216,* 18–24.

12. Mukesh, D. J., Balakumaran, M. D., Kalaichelvan, P. T., Pandey, A., Singh, A., & Raja, R. B., (2011). Isolation, production & application of extracellular phytase by *Serratia marcescens. J. Exp. Biol. Sci., 2,* 663–666.

13. Shimizu, M., (2014). Purification and characterization of phytase from *Bacillus suhtilis* (natto) N–77. *Biosci. Biotechnol. Biochem., 56,* 1266–1269.

14. Kerovuo, J., Lauraeus, M., Nurminen, P., Kalkkinen, N., & Apajalahti, J., (1998). Isolation, characterization, molecular gene cloning, and sequencing of a novel phytase from *Bacillus subtilis. Appl. Environ. Microbiol., 64,* 2079–2085.

15. Kerovuo, J., Lappalainen, I., & Reinikainen, T., (2000). The metal dependence of *Bacillus subtilis* phytase. *Biochem. Biophys. Res. Commun., 268,* 365–369.

16. Sunitha, K., Lee, J., & Oh, T., (1999). Optimization of medium components for phytase production by *E. coli* using response surface methodology. *Bioproc. Biosystems. Eng., 21,* 477–481.

17. Lim, D., Golovan, S., Forsberg, C. W., & Jia, Z., (2000). Crystal structures of *Escherichia coli* phytase and its complex with phytate. *Nat. Struct. Mol. Biol., 7,* 108–113.

18. Sreeramulu, G., Srinivasa, D. S., Nand, K., & Joseph, R., (1996). *Lactobacillus amylovorus* as a phytase producer in submerged culture. *Lett. Appl. Microbiol., 23,* 385–388.

19. Richardson, A. E., & Hadobas, P. A., (1997). Soil isolates of *Pseudomonas* spp. that utilize inositol phosphates. *Can. J. Microbiol., 43,* 509–516.

20. Yanke, L. J., Bae, H. D., Selinger, L. B., & Cheng, K. J., (1998). Phytase activity of anaerobic ruminal bacteria. *Micro. Soc., 144,* 1565–1573.

21. Yanke, L. J., Selinger, L. B., & Cheng, K. J., (2002). Phytase activity of *Selenomonas ruminantium*: A preliminary characterization. *Lett. Appl. Microbiol., 29,* 20–25.

22. Segueilha, L., Moulin, G., & Galzy, P., (1993). Reduction of phytate content in wheat bran and glandless cotton flour by *Schwanniomyces castellii. J. Agric. Food Chem., 41,* 2451–2454.

23. Nakamura, Y., Fukuhara, H., & Sano, K., (2000). Secreted phytase activities of yeasts. *Biosci. Biotechnol. Biochem., 64,* 841–844.

24. Kim, D., Godber, J. S., & Kim, H., (1999). Culture conditions for a new phytase-producing fungus. *Biotechnol. Lett., 21,* 1077–1081.

25. Costa, M., Torres, M., Magariños, H., & Reyes, A., (2010). Production and partial purification of *Aspergillus ficuum* hydrolytic enzymes in solid state fermentation of agro-industrial residues. *Rev. Colomb. Biotecnol., 12,* 163–175.

26. Papagianni, M., Nokes, S. E., & Filer, K., (2001). Submerged and solid-state phytase fermentation by *Aspergillus niger*: Effects of agitation and medium viscosity on phytase production, fungal morphology and inoculums performance. *Food. Technol. Biotech., 39,* 319–326.

27. Almeida, F. N., Sulabo, R. C., & Stein, H. H., (2013). Effects of a novel bacterial phytase expressed in *Aspergillus oryzae* on digestibility of calcium and phosphorus in diets fed to weanling or growing pigs. *J. Animal. Sci. Biotechnol., 4,* 1–10.

28. El Gindy, A. A., Ibrahim, Z. M., Ali, U. F., & El Mahdy, O. M., (2009). Extracellular phytase production by solid-state cultures of *Malbranchea sulfurea* and *Aspergillus niveus* on cost-effective MEDIUM. *J. Agric. Biol. Sci., 5,* 42–62.

29. Haefner, S., Knietsch, A., Scholten, E., Braun, J., Lohscheidt, M., & Zelder, O., (2005). Biotechnological production and applications of phytases. *Appl. Microbiol. Biotechnol., 68,* 588–597.

30. Marlida, Y., Delfita, R., Gusmanizar, N., & Ciptaan, G., (2010). Identification characterization and production of phytase from endophytic fungi. *World Acad. Sci. Eng. Technol., 65,* 1043–1046.

31. Coban, H. B., Demirci, A., & Turhan, I., (2015). Enhanced *Aspergillus ficuum* phytase production in fed-batch and continuous fermentations in the presence of talcum microparticles. *Bioprocess Biosyst. Eng., 38,* 1431–1436.

32. Bogar, B., Szakacs, G., Linden, J. C., Pandey, A., & Tengerdy, R. P., (2003). Optimization of phytase production by solid substrate fermentation. *J. Ind. Microbiol. Biotechnol., 30,* 183–9.

33. Franco, I., (2007). Optimización de la producción de fitasa por *Aspergillus niger* en Fermentación en Estado Sólido utilizando métodos estadísticos. MSc dissertation, Instituto Politecnico Nacional, México.

34. Mittal, A., Singh, G., Goyal, V., Anita, Y., Aggarwal, N. K., & Kumar, N., (2011). optimization of medium components for phytase production on orange peel flour by *Klebsiella* Sp. DB3 using response surface methodology. *Inov. Romnian Food Biotechnol., 9*, 35–44.

35. Ojeda, A., Frías, A., González, R., & Linares, Z., (2010). Tannins, phytic phosphorus, phytase activity in the seed of 12 sorghum grain hybrids (*Sorghum bicolor* (L) Moench). *Arch. Latinoam. Nutr., 60*, 1–6.

36. Godoy, S., & Chico, C. F., (2005). Phytate phosphorus utilization in ruminant nutrition. *Rev. Digit. Ceniap., 9*, 1–7.

37. Romero, C., Salas, M., García, A. C., Mendoza, G., Plata, F., Cervantes, M., Viana, T., & Morales, A., (2009). Effect of phytase from *Aspergillus niger* on nutrient digestibility and activity of trypsin and chymotrypsin in weanling pigs. *Arch. Zootec., 58*, 363–369.

38. Sangronis, E., Torres, A., & Sanabria, N., (2006). Phytases and α-galactosides in germinated seeds of *Phaseolus vulgaris* and *Vigna sinesis*. *Agron. Trop., 56*, 523–529.

39. McKevith, B., (2004). Nutritional aspects of cereals. *Nutr. Bull., 29*, 111–142.

40. Morales, G. A., Moyano, F. J., & Marquez, L., (2011). *In vitro* assessment of the effects of phytate and phytase on nitrogen and phosphorus bioaccessibility within fish digestive tract. *Anim. Feed Sci. Technol., 170*, 209–221. doi: 10.1016/j.anifeedsci.2011.08.011.

41. Cuca, M., De La Rosa, G., Pró, A., & Baeza, J., (2003). Phosphorus availability in soybean meal and sorghum-corn gluten meal, supplemented with phytase for starter broilers. *Tec. Pecu. Méx., 41*, 295–306.

42. Hernández, G., Godoy, S., & Chicco, C., (2005). Phytases activity and phosphorus absorption from cereals in chicks. *Rev. Cient. Univ. Zulia., XV*, 505–511.

43. Sotelo, A., Mendoza, J., & Argote, R., (2002). Contenido de ácido fítico en algunos alimentos crudos y procesados. Validación de un método colorimétrico. *J. Mex. Chem. Soc., 46*, 301–306.

44. Rivera, J. G., Peraza, F. A., Serratos, J. C., Posos, P., Guzmán, S. H., Cortez, E., Castañón, G., & Mendoza, M., (2009). Effect of nitrogen and phosphorus fertilization on phytic acid concentration and vigor of oat seed (var. Saia) in Mexico. *J. Exp. Bot., 78*, 37–42.

45. Cortés, A., Fuentes, B., Fernández, S., Mojica, M., & Ávila, E., (2007). Evaluation of the presence of a microbial phytase (*Peniophora lycii*) in sorghum-soybean phosphorus-deficient meal diets, for broilers, on ileal protein, and amino acid digestibility and metabolizable energy. *Vet. México., 38*, 21–30.

46. Eklund, C., Sandberg, A., & Larsson, M., (2006). Reduction of phytate content while preserving minerals during whole grain cereal Tempe fermentation. *J. Cereal Sci., 44*, 154–160.

47. Anderson, P. A., (1987). Digestibility and amino acid availability in cereals and oilseeds. *Hopkins. Am. Assoc. Cereal Chem.*, 31–46.

48. Kumar, V., Sinha, A. K., Makkar, H. P. S., & Becker, K., (2010). Dietary roles of phytate and phytase in human nutrition: A review. *Food Chem., 120*, 945–959.

49. Noureddini, H., & Dang, J., (2010). An integrated approach to the degradation of phytates in the corn wet milling process. *Bioresour. Technol., 101*, 9106–9113.
50. Spier, M. R., Greiner, R., Rodriguez-León, J. A., & Lorenci, A., (2008). Phytase production using citric pulp and other residues of the agroindustry in SSF by fungal isolates. *Food Technol. Biotechnol.,* 178–182.
51. Bertinato, J., Sherrard, L., & Plouffe, L. J., (2012). EDTA disodium zinc has superior bioavailability compared to common inorganic or chelated zinc compounds in rats fed a high phytic acid diet. *J., Trace Elem. Med. Biol., 26*, 227–233.
52. Bilgiçli, N., Elgün, A., & Türker, S., (2006). Effects of various phytase sources on phytic acid content, mineral extractability and protein digestibility of tarhana. *Food Chem., 98*, 329–337.
53. Reis, M., & Aparecida, M., (1999). Nutritional aspects of phytates and tannins. *Rev. Nutr. Campinas, 12*, 21–32.
54. Neira, A., Nava, E., Lozano, J., Michelena, G., Gaona, J., Martínez, J., & Iliná, A., (2011). Evaluation of antinutritional factor phytic acid on triticale AN66 from Coahuila State (Mexico), at different phenological state. In: García, S., García-Galindo, H. S., & Nevárez-Morillón, G. V., (eds.), *Advances in Science, Biotechnology and Safety of Foods*. Asociación Mexicana de Ciencias de los Alimentos A. C., México.
55. Cortés, F. J., Turrent, F., Díaz, V., Jiménez, S., Hernández, R., & Mendoza, R., (2005). Hillside agriculture and food security in Mexico Advances in the sustainable hillside management project. In: Lal, R., Stewart, B. A., Uphoff, N., & Hansen, D. O., (eds.), *Climate and Global Food Security* (pp. 569–588). Tylor and Francis: Boca Raton, Florida.
56. Singh, B., & Satyanarayana, T., (2011). Phytases from thermophilic molds: Their production, characteristics and multifarious applications. *Process Biochem., 46*, 1391–1398.
57. Kim, E. Y., Kim, Y., Rhee, M. H., Song, J., Lee, K., Kim, K., Lee, S. P., Lee, I., & Park, S. C., (2007). Selection of *Lactobacillus* sp. PSC101 that produces active dietary enzymes such as amylase, lipase, phytase and protease in pigs. *J. Gen. Appl. Microbiol., 53*, 111–117.
58. Cervantes, M., Sauer, W. C., Morales, A., Araiza, B., Espinoza, S., & Yánez, J., (2009). Nutritional manipulation and pollution. *Rev. Comput. Prod. Porc., 16*, 13–22.
59. Frontela, C., Ros, G., & Martínez, C., (2008b). Application of phytases as functional ingredient in foods. *Arch. Latinoam. Nutr., 58*, 215–220.
60. Ramesh, K. P., & Prakash, V., (2009). The stabilizing effects of polyols and sugars on porcine pancreatic lipase. *J. Am. Oil Chem. Soc., 86*, 773–781.
61. Ravindran, V., Cabahug, S., Ravindran, G., & Bryden, W. L., (1999). Influence of microbial phytase on apparent ileal amino acid digestibility of feedstuffs for broilers. *Poult. Sci., 78*, 699–706.
62. Harbach, P., Da Costa, M., Soares, A., Bridi, A., Shimokomaki, M., Da Silva, C., & Ida, E., (2007). Dietary corn germ containing phytic acid prevents pork meat lipid oxidation while maintaining normal animal growth performance. *Food Chem., 100*, 1630–1633.
63. Albarracín, M., González, R. J., & Drago, S. R., (2013). Effect of soaking process on nutrient bio-accessibility and phytic acid content of brown rice cultivar. *LWT - Food Sci. Technol., 53*, 76–80.

64. Rebollar, P. G., & Mateos, G. G., (1999). El fósforo en nutrición animal. necesidades, valoración de materias primas y mejora de la disponibilidad. In: Rebollar, P. G., Beorlegui, C., & González, M., (eds.), *Avances en Nutrición Animal y Alimentación Animal* (pp. 19–64). Madrid, España.

65. Menezes-Blackburn, D., Jorquera, M., Gianfreda, L., Rao, M., Greiner, R., Garrido, E., De La Luz, M. M., et al., (2011). Activity stabilization of *Aspergillus niger* and *Escherichia coli* phytases immobilized on allophanic synthetic compounds and montmorillonite nanoclays. *Bioresour. Technol., 102*, 9360–7.

66. Mandviwala, T. N., & Khire, J. M. M., (2000). Production of high activity thermostable phytase from thermotolerant *Aspergillus niger* in solid state fermentation. *J. Ind. Microbiol. Biotechnol., 24*, 237–243.

67. Rutherfurd, S. M., Chung, T. K., Morel, P. C. H., & Moughan, P. J., (2004). Effect of microbial phytase on ileal digestibility of phytate phosphorus, total phosphorus, and amino acids in a low-phosphorus diet for broilers. *Poult. Sci., 83*, 61–68.

68. Frontela, C., Haro, J. F., Ros, G., & Martinez, C., (2008a). Effect of dephytinization and follow-on formula addition on *in vitro iron*, calcium, and zinc availability from infant cereals. *J. Agric. Food Chem., 56*, 3805–38011.

69. Ali, H. K. Q., & Zulkali, M. M. D., (2011). Utilization of agro-residual ligno-cellulosic sub- stances by using solid state fermentation : A review. *Croat. J. Food Technol. Biotechnol. Nutr., 6*, 5–12.

70. Liu, B., Rafiq, A., Tzeng, Y., & Rob, A., (1998). The induction and characterization of phytase and beyond. *Enzyme Microb. Technol., 229*, 415–424.

71. Bhavsar, K., Ravi, V., & Khire, J. M., (2012). Downstream processing of extracellular phytase from *Aspergillus niger*: Chromatography process vs. aqueous two phase extraction for its simultaneous partitioning and purification. *Process Biochem., 47*, 1066–1072.

72. Zou, L. K., Wang, H. N., Pan, X., Tian, G. B., Xie, Z. W., Wu, Q., Chen, H., Xie, T., & Yang, Z. R., (2008). Expression, purification and characterization of a phyA(m)-phyCs fusion phytase. *J. Zhejiang Univ. Sci. B., 9*, 536–545.

73. Krishna, C., & Nokes, S. (2001). Predicting vegetative inoculum performance to maximize phytase production in solid-state fermentation using response surface methodology. *J. Ind. Microbiol. Biotechnol., 26*, 161–170.

74. Papagianni, M., Nokes, S. E., & Filer, K., (1999). Production of phytase by *Aspergillus niger* in submerged and solid-state fermentation. *Process Biochem., 35*, 397–402.

75. Rodríguez, D. E., Rodríguez, J. A., Carvalho, J. C., Thomaz, V., Parada, J. L., & Soccol, C. R., (2010). Recovery of phytase produced by solid-state fermentation on citrus peel. *Brazilian Arch. Biol. Technol. an Int. J., 53*, 1487–1496.

76. Fujita, J. I. N., Shigeta, S., Yamane, Y., Fukuda, H., Kizaki, Y., Wakabayashi, S., & Ono, K., (2003). Production of two types of phytase from *Aspergillus oryzae* during industrial koji making. *J. Biosci. Bioeng., 95*, 460–465.

77. Omogbenigun, F. O., Nyachoti, C. M., & Slominski, B. A., (2003). The effect of supplementing microbial phytase and organic acids to a corn-soybean based diet fed to early-weaned pigs. *J. Anim. Sci., 81*, 1806–1813.

78. Pointillart, A., Fourdin, A., & Fontaine, N., (1986). Importance of cereal phytase activity for phytate phosphorus utilization by growing pigs fed diets containing triticale or corn. *J. Nutr., 117*, 907–913.

79. Konietzny, U., & Greiner, R., (2004). Bacterial phytase: Potential application, *in vivo* function and regulation of its synthesis. *Brazilian J. Microbiol., 35*, 11–18.

80. Quan, C., Zhang, L., Wang, Y., & Ohta, Y., (2001). Production of phytase in a low phosphate medium by a novel yeast *Candida krusei*. *J. Biosci. Bioeng., 92*, 154–160.

81. Bergman, E., Autio, K., & Sandberg, A., (2000). Optimal conditions for phytate degradation, estimation of phytase activity, and localization of phytate in barley (cv. Blenheim). *J. Agric. Food Chem., 48*, 4647–4655.

82. Ribeiro, C. T. L., De Queiroz, M. V., & De Araújo, E. F., (2015). Cloning, recombinant expression and characterization of a new phytase from *Penicillium chrysogenum*. *Microbiol. Res., 170*, 205–212.

83. Fei, B., Xu, H., Zhang, F., Li, X., Ma, S., Cao, Y. Y., Xie, J., Qiao, D., & Cao, Y. Y., (2013). Relationship between *Escherichia coli* AppA phytase's thermostability and salt bridges. *J. Biosci. Bioeng., 115*, 623–7.

84. Gaind, S., & Singh, S., (2015). Production, purification and characterization of neutral phytase from thermotolerant *Aspergillus flavus* ITCC 6720. *Int. Biodeterior. Biodegradation, 99*, 15–22.

85. Díaz, A. B., Caro, I., De Ory, I., & Blandino, A., (2007). Evaluation of the conditions for the extraction of hydrolitic enzymes obtained by solid state fermentation from grape pomace. *Enzyme Microb. Technol., 41*, 302–306.

86. Grimm, L. H., Kelly, S., Krull, R., & Hempel, D. C., (2005). Morphology and productivity of filamentous fungi. *Appl. Microbiol. Biotechnol., 69*, 375–384.

87. Saad, N., Esa, N. M., Ithnin, H., & Shafie, N. H., (2011). Optimization of optimum condition for phytic acid extraction from rice bran. *African J. Plant Sci., 5*, 168–175.

88. AOAC, (1990). *Official Methods of Analysis*, 15th ed. Association of Analytical Chemistry, Washington, DC.

89. El-Gindy, A. A., Ibrahim, Z. M., Ali, U. F., & El-Mahdy. (2009). Extracellular phytase production by solid-state cultures of *Malbranchea sulfurea* and *Aspergillus niveus* on cost-effective medium. *Res. J. Agric. Biol. Sci., 5*, 42–62.

90. Bhavsar, K., Kumar, V. R., & Khire, J. M. (2011). High level phytase production by *Aspergillus niger* NCIM 563 in solid state culture: response surface optimization, up-scaling, and its partial characterization. *J. Ind. Microbiol. Biotechnol., 38*, 1407–1417.

CHAPTER 11

Advances in the Biotechnological Process for Obtaining Ellagic Acid from Rambutan

NADIA D. CERDA-CEJUDO,[1] JOSE J. BUENROSTRO-FIGUEROA,[2] LEONARDO SEPÚLVEDA,[1] CRISTIAN TORRES-LEÓN,[3] MÓNICA L. CHÁVEZ-GONZÁLEZ,[1] CRISTÓBAL NOÉ AGUILAR,[1] and J. A. ASCACIO-VALDÉS[1]

[1]*Bioprocesses and Bioproducts Research Group, Food Research Department, School of Chemistry, Autonomous University of Coahuila, Saltillo – 25280, Coahuila, México,*
E-mail: alberto_ascaciovaldes@uadec.edu.mx (J. A. Ascacio-Valdés)

[2]*Research Center for Food and Development A.C., Cd. Delicias – 33088, Chihuahua, México*

[3]*Research Center and Ethnobiological Garden, Autonomous University of Coahuila, Viesca, Coahuila, México*

ABSTRACT

Polyphenols are secondary metabolites distributed in the kingdom Plantae, as part of the bark, leaves, stems, flowers, and fruits; these polyphenols have been currently classified according to their structural characteristics into flavonoids, condensed tannins, phlorotannins, hydroxystilbenes, and hydrolyzed tannins [1]. Within the latter group are the ellagitannins which are esters od hexahydroxydiphenic acid (HHDP) usually linked to glucose, so when the esters link is hydrolyzed by enzymes or acid-basic methods, the HHDP group is released and undergoes molecular arrangement to create ellagic acid [2].

Ellagic acid is a hydrolysable polyphenol derived from the biotransformation of some biocompounds, like ellagitannase and corilagine present in fruits like rambutan, pomegranate, red fruits, and nuts. The bioactive compound has been used in several industries like pharmaceutical, cosmetics, and food industries for the benefits that provide to human health. This molecule has antioxidant, antiviral, antitumoral, and anticancerogenic activities.

The biotechnological production of ellagic acid is a feasible alternative like getting it from microbiological sources, in which the fungus strains have demonstrated that are the best to biotransform the compounds of the sources to get ellagic acid. In order to obtain ellagic acid, the extraction assisted from solid-state fermentation (SSF) has been an interesting way to obtain this molecule. This technology is practical, innovative, and inexpensive so it can be implemented in developing countries.

This chapter is focus in the rambutan fruit as source of ellagitannins, mainly, ellagic acid that rambutan fruit can provide and purpose the biotechnological extraction method to obtain this biocompound for the advantages that this bioprocess has above conventional methods and emergent technologies. This chapter pretends to establish a new perspective in the food industry.

11.1   INTRODUCTION

Natural polyphenols have an essential group, the tannins, which are divided into condensed tannins and hydrolysable tannins [3]. Hydrolyzable tannins have two subclasses: the ellagitannins and the gallotannins [4].

The ellagitannins are polyphenolic compounds which are characterized by the presence of units of gallotannins and hexahydroxydiphenoyl (HHDP) so called biarylic dehydrodigalloyl [4], which are esters with a molecular weight of 300 to 20,000 Da. Ellagitannins are mostly water-soluble, they act like weak acids and they can be found in plants where they are among the most abundant groups of polyphenolic compounds of their secondary, they accumulate in vacuoles of the cells and act as protection of the plant in a stress situation, in addition to helping in some functions of survival and adaptation of the plants to several environments [5].

The ellagitannins are composed of units of gallic acid attached to a central sugar core, which is glucose and contains units of gallic acid linked to each other by steric bonds of C-C baryl and diaryl ether CO, resulting from the oxidative coupling of the phenolic compounds intra and inter molecular [6].

The HDDP group that is the monomeric unit of the ellagitannins, due to the oxidation process mentioned above and it is bound to polyols by ester bonds [7]. When hydrolysis of ellagitannins occurs, the HDDP groups separate from the glycosidic nucleus, and since these compounds are not stable, they are lactonized quickly leading to the formation of a dilactone, ellagic acid [8]. In Figure 11.1, is shown the synthesis of ellagitannins proposed by Chen and Hagerman [9]; and Chávez et al. [10], where the molecule of geraniin is lactonized and lose its HDDP group and the glucose group to create the ellagic acid.

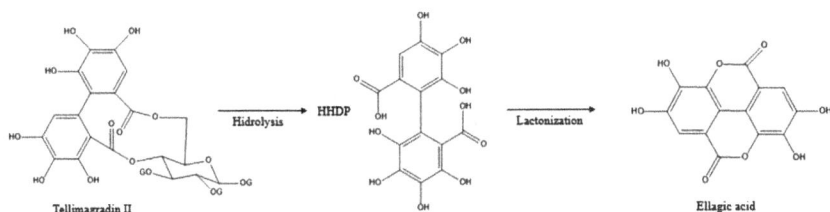

FIGURE 11.1   Synthesis of ellagitannins.

11.2   ELLAGIC ACID

Ellagic acid is a dilactone and potent bioactive compound; it is a polyphenol that can be found in pomegranate, rambutan, grapes, walnut, and berries. It is widely used in several industries such as food in which is consumed in jams, fruit juices and other beverages [11], pharmaceutical industry in which some studies have demonstrated that this ellagitannin has antiepileptic activity in mice, this is probably through the increase of GABAering transmission in brain [12], also the ellagic acid with the inhibition of aldorecutase activity, has a therapeutic promise to treat or prevent complications of diabetes by this ellagitannin present in many dietary sources, and cosmetics. Ellagic acid is an important molecule due to its biological activities, as antimicrobial, antiproliferative, antitumoral, and antiviral, promotes good in human's health (Figure 11.2) [13].

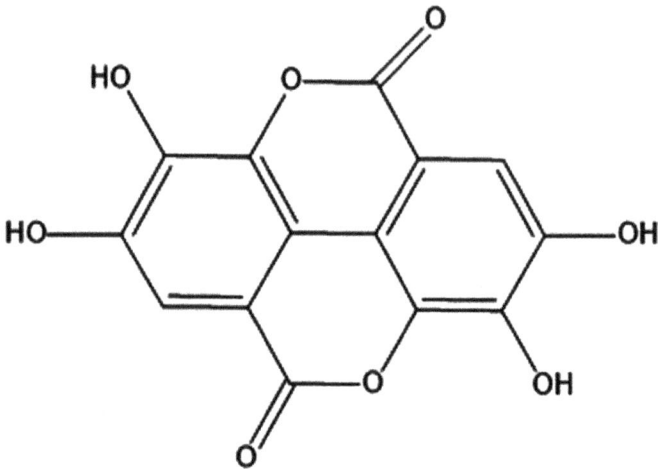

FIGURE 11.2   Chemical structure of ellagic acid.

Most of the ellagic acid produce by conventional methods is produced by acid hydrolysis and solvent extraction, which can be polluting with the environment, and also, these techniques can be harmful in the recovery and purification of this compound, also, the cost is increased. Recently by exploiting agro-industrial wastes using fungal strains and fermentation processes, the ellagic acid production has been carried out [14].

Sepúlveda et al. [15] separated ellagic acid from pomegranate produced by a SSF process using *Aspergillus niger GH1* purifying it by molecular-size exclusion chromatography and subsequent identification using HPL/ESI/MS; being able to separate the ellagic acid. In pomegranate the main polyphenol compounds were punicalin, punicalagin, and ellagic acid which demonstrated that is a good alternative to use this bioprocess for purification and separation of ellagic acid, which has already been mention, can be applied in food science or in the pharmaceutical industry.

Likewise, another research study describes the accumulation of ellagic acid by a SSF using *A. niger* GH1 and powdered pomegranate peel as support [2]; authors evaluated several culture conditions, such as temperature, initial humidity, inoculum levels, and concentrations of salts, being the temperature, the concentration of salts of potassium chloride and magnesium sulfate, the significant parameters in the accumulation of ellagic acid.

11.3 SOURCES OF ELLAGITANNINS

Ellagitannins can be found in several variety of plants (Table 11.1), such as oaks or chestnuts and some fruits like pomegranate, rambutan, grapes, walnut, and red fruits, like blueberries and raspberries, mainly through the quantification of the ellagic acid present [16].

TABLE 11.1 Sources of Ellagitannins

Source	Ellagitannins	References
Pomegranate	Punicalin, punicalagin	[17, 18]
Raspberry	Penduculagine, sanguine, Galloylbis-HHDP glucose isomer	[19]
Wine (oak)	Vascalagine	[20]
Rambután	Geranine, corilagine, ellagic acid	[21]
Blackberry	Penduculagine, castalagine/vascalagine, lambertianine	[22]
Nuts	Ellagic acid	[23]

11.4 RAMBUTÁN

Rambutan is a tropical fruit, belongs to *Sapindaceae* family original from the southeast region of the Asian continent, mainly from the countries of Malaysia, Indonesia, and Thailand [24]. It is defined as an exotic fruit that began to be cultivated in Central America where the agroclimatic conditions are similar to the conditions of origin.

The name of the rambutan derives from the Malay Word "rambut" which means hair, this to the fact that the pericarp of the fruit is covered with a kind of soft spines that resemble hair [25]. The rambutan tree is medium size and its normally 12–20 meters tall, its leaves are perennial, 10 to 30 cm long with a color that ranges from light to dark green, the leaves are made up of 3 to 11 leaflets, 5 to 15 cm long and 3 to 10 cm long [26, 27].

Rambutan fruit is a drupe that measures around 2 to 3 cm and is made up of peel, pulp, and seed, the shell is between 3 and 4 cm thick and has a range of colors that can vary from yellow, red, brown, and orange; rambutan is characterized by the presence of soft spines around the entire fruit (Figure 11.3) [28].

FIGURE 11.3   Rambutan: Peel, pulp, and seed.

The pulp covers the seed which has an almond-like shape and is approximately 2 cm wide and 3 cm long. In general, the fruit has a sweet flavor, but sometimes it can be a little acidic; the harvest of the fruit is carried out once it has reached its ripening, which is distinguished by the color range in the fruit [28].

The pulp of the fruit is normally consumed fresh. Industrially, it is used in several products, such as juices, jellies, or jams [29]. The pulpis the only edible part of the fruit, so the peel and seed are considered residues, which can be used for the development of alternatives for use and recovery due to a several compounds present in them, and evaluate their possible biological properties [30].

Solís-Fuentes et al. [31] reported that rambutan fruit has pulp, seed, and peel, with a weight average formed by all of its parts which consisted in the ones that already been mentioned, and the percentages by weight of each part of the fruit is represented in Table 11.2.

TABLE 11.2   Percentage and Weight of Constituent Portions in Rambutan Fruit

Fruit Part	Proportion (%)	Weight (g)
Whole fruit	100	27.4 (±2.2)
Seed	9.5 (±0.7)	2.53 (±0.22)
Pulp	44.8 (±2.5)	11.7 (±0.6)
Peel	45.7 (±3.2)	13.2 (±1.8)
Embryo	6.1 (±0.6)	1.60 (±0.21)

11.4.1   SEED

For the rambutan seed, several researches have been carried out on the physicochemical properties, in which the seed has a weight of approximately 6.1% of the fruit. In a proximal analysis determined by Serida et al. [32], the data obtained were that the seed possess 12.4% of protein, 48% of carbohydrates, 2.26% ashes and 3.31% of moisture; also, the main fatty acids evaluated of the rambutan seed were 40.45% of oleic acid and 36.36% of arachidonic acid.

Mahmood et al. [27] reported that abundant fat content above 14–41% with oleic acid, renders the seed a novel source of vegetable fat; the authors considered the possibilities of using seed fat in chocolate (30 wt.% substitute) and personal care products are also on the focus. Authors also mentioned the nanostructured seed fat is reported for encapsulation off soluble vitamins (like vitamin E) and additionally, the seed contains most of the essential and non-essential amino acid that are concentrated as protein concentrate.

Due to their high content of nutritional compounds such as proteins, oil, and carbohydrates among others, rambutan seeds can be used in food industry for several applications.

11.4.2   PULP

According to Fraire [33], the pulp of the rambutan is consumed fresh, due to it is the only edible part of the fruit and water is the largest component. Fila et al. [34] determined the physicochemical analysis of rambutan pulp (based on (g/100 g) in which the proximal composition of fresh rambutan was: moisture (78.46 g), protein (0.66 g), crude fiber (0.38 g), amino acids (19.66 g), carbohydrates (0.24 g), and ash (0.60 g). Compared to the rambutan seed, the fat content in the pulp is lower [35].

Vargas [36] reported a study with the micro and macronutrients of rambutan pulp in which nitrogen was the most abundant macronutrient in the fruit while sulfur was the least, the values of the nutrients reported are shown in Table 11.3.

TABLE 11.3 Content of Minerals in Rambutan Pulp

Minerals in Rambutan Pulp (mg) in Dry Base					
Macronutrients	**S**	**Mg**	**P**	**Ca**	**K**	**N**
	4–6	9–13	11–13	22–31	63–81	77–87
Micronutrients	**Cu**	**Zn**	**B**	**Fe**	**Mn**	
	0.08–0.10	0.09–0.11	0.12–0.16	0.16–0.23	0.26–0.38	−

11.4.3 PEEL

The rambutan peel is one of the main constituents of the fruit, which can reach up to 43–57% of total weight depending on the type of crop and its maturation. Mahmood et al. [27] determined the proximal composition of rambutan peel (based on (g/100 g) in which the physicochemical analysis of rambutan peel: moisture (72.05), lipids (0.23), carbohydrates (23.78), protein (2.04), fiber (0.7) and ashes (1.2).

Also, Hernández et al. [37] determined and reported the mineral content of dry rambutan peel in mg/L with the following results: Mn (0.14), Cu (0.070), Na (0.04), Fe (0.29), Zn (0.080), Mg (0.15) and Ca (0.51); also exhibits a chemical composition of fiber as cellulose with a reported value of 24.28 ± 2.30 (% w/w), hemicellulose with a reported content of 11.62 ± 2.31 (% w/w) and lignin with a reported value of 35.34 ± 2.05 (% w/w) [38].

In Table 11.4, we can see the values of the proximal analysis of the seed, pulp, and peel of rambutan fruit. In the physicochemical analysis of fiber, it can be compared that the peel has a higher value over the pulp, the authors did not report realizing the fiber analysis of the seed, but instead, they reported the fatty acids that the seed possess (like oleic acid). Also, it can be compared the moisture of each part of the fruit, being the pulp with the highest value.

TABLE 11.4 Physicochemical Analysis of Rambutan Fruit

Compound	Seed (%) [32]	Pulp (g/100 g) [34]	Peel (g/100 g) [29]
Moisture	3.3	78.4	72.0
Lipids	−	19.6	0.2
Carbohydrates	48.0	0.2	23.7

TABLE 11.4 *(Continued)*

Compound	Seed (%) [32]	Pulp (g/100 g) [34]	Peel (g/100 g) [29]
Protein	12.4	0.6	2.0
Fiber	–	0.3	0.7
Ashes	2.2	0.6	1.2
Oleic acid	40.4	–	–
Arachidonic acid	36.3	–	–

Rambutan peel has a high content of phenolic compounds that have been reported by Okonogi et al. [39]; and Palanisamy et al. [40], in their researches they also reported it has a high antioxidant activity, also, the rambutan peel has been reported that it is a potential source of antioxidants to be implemented in pharmaceuticals, cosmetics, and food products due to its nontoxic capacity for normal cells and antioxidant activity content. The peel is an interesting source of bioactive compounds compared to rambutan seed or pulp extracts, the amount of antioxidants compounds is greater in which it contains a greater amount of antioxidants compounds, which are mainly polyphenolic compounds in which hydrolyzed compounds are the greater amount, known for their several biological activities which includes corylagine, ellagic acid and geraniin, these are a group of ellagitannins that are present in rambutan peel, also, geraniin is the major compound present [41].

According to Škerget et al. [42], one factor that must be considered to known the polyphenol content in rambutan peel is the cultivation and fruit stage development. However, some factors have been studied to obtains these compounds such as extraction technique, type of solvent, temperature, particle, size, pH, and solvent-to-solid ratio, which contribute to the efficacy of polyphenol extraction [43].

Due to the presence of these compounds and their importance derived from the attributed biological activities, the interest in taking advantage of agro-industrial waste, such as rambutan peel, has increased. It has been reported that rambutan peel extracts due to the presence of phenolic acid compounds and ellagitannins, have a high antioxidant activity due [21].

Palanisamy et al. [40] evaluated the antioxidant activity of an etha-nolysis extract of rambutan peels compared to grape extract and vitamin C, the result of this research showed the extract of leaves and rind of rambutan fruit displayed the highest capacity to inhibit the radical HHDP with the ethanolic rind extracts having the highest $1/IC_{50}$ value compared to a grape extract and vitamin C.

The antimicrobial activity has been reported for extracts of rambutan peel. Sektar et al. [68] showed that bacteria Gram (+) had been sensi-tive to the actions of the methanolic compounds of rambutan peel. In another research, inhibitory effect of the growth of microorganisms (pathogenic bacteria like *Staphylococcus epidermidis, Enterococcus faecalis,* and *Staphylococcus aureus*) was observed using extracts of rambutan [69].

Several anticancer studies have shown the antiproliferative activity of the extracts of rambutan peel against MDA-MB-231 and MG-63 cell lines for breast cancer with IC_{50} values of 5.42 ±1.67 µ/mL, HeLa cell lines for cervical cancer and MG-63 osteosarcoma cancer cell lines [70].

Another study demonstrated the anti-inflammatory and analgesic activity of methanol extract from rambutan seeds has. The biological activities (anti-inflammatory and analgesic) of 58.86% and 51.27% respectively were exhibited from the extract [71]. Also, studies performed by Ahmad et al. [44] with rambutan peel in which were observed antiviral activity by inhibiting the mechanism of viral choring against the type 2 dengue virus (DENV-2), with an IC_{50} of 1.75 µM, besides, the studies demonstrated that the compound inhibits viral binding by binding to the E-DIII protein and interfering with the initial cell-virus interaction.

11.5   EXTRACTION OF ELLAGITANNINS

The ellagitannins have been extracted mainly by conventional methods and recently the use of new technologies has taken a boom. In conventional methods, organic solvents are used such as methanol, used water, acetone, ethanol, and ethyl acetate in order of use [45]. Hernández et al. [37]; and Thitilertdecha et al. [21] reported that ellagitannins can be obtained by a methanolic or ethanolic respectively.

In order to guarantee the extraction efficiency of ellagitannins on the extraction process, a sample pretreatment before the extraction may be needed, also, the sample pretreatment depends to the moisture content and the complexity of the sample matrix, which, depending on the case, may decrease the extraction of the analyte from inside the cell, if the sample gets dried [46]. Due to the water content, in some cases, the stability of the sample could decrease because of enzymatic reactions that can happen due to the presence of water [47].

The use of new technologies is looking to reduce the use of contaminant solvents and getting better results in the extractions of these components. Some of the latest technologies are pressurized liquid extraction, ultrasound (US)-assisted extraction, simultaneous distillation, pressurized extraction with hot water, microwave-assisted extraction, and supercritical fluids extraction [45].

11.6   EMERGENT EXTRACTION TECHNOLOGIES

Through the years, new technologies have emerged as alternatives to standard extractions, with more significant advantages, because they are automated, with higher reproducibility and selectivity and they are faster than conventional methods. Furthermore, the solvents used in these technologies are less polluting and more sustainable than the solutions used in traditional technologies [48].

These techniques include simultaneous distillation dual, pressurized liquid extraction, US-assisted extraction, pressurized extraction with hot water, microwave-assisted extraction, and supercritical fluid extraction [72]. However, one of the disadvantages of these techniques is the complexity and difficulty to be implemented in developing countries, as well as the use of high temperatures which can damage the compounds. Currently, the use of other sustainable techniques for tannin extraction has been requested, so the bioprocess being one of them. Among the bioprocess used for these purposes, SSF stands out.

11.7   SOLID-STATE FERMENTATION (SSF)

According to Cannel and Moo-Young [49], SSF is defined as "the fermentative process in which microorganisms grow in solid materials without the presence of free liquid." This may be possible because of the moisture necessary for the growth of the microorganisms in the substrate.

Two types of SSFs have been distinguished; in the first one, the microorganism grows in the natural material (like agro-industrial waste), and in the second, the substrate where the microorganism grows is inert, but this is impregnated with liquid culture medium to promote the growth [50].

Since it requires less capital, small equipment, and lower operating cost, developing this biotechnological process seems easy to implement, in addition to having environmental advantages, since it allows the use of agro-industrial waste as a substrate [51]. Also, we must consider the disadvantages that this process has, like the microorganism used, which has to be one that grows at low water content (like filamentous fungus) because one of the factors of SSF is the low levels of water that this process has, the reactors used (like Petri dish in the laboratory) and to scale it to be used in industrial range and the time used in developing this bioprocess for obtaining the compound in interest.

The influence of inert supports of the production of ellagic acid has not been studied at all, therefore, the study was carried out by this method with the *Aspergillus niger GH1* strain using three inert supports of polyurethane foam, nylon fiber and perlite using as carbon source partially purified pomegranate peel ellagitannins [52]. The authors obtained a high yield in the production of ellagic acid in a shorter period compared to emergent techniques of extraction of phenolic compounds [53] providing an alternative to produce ellagic acid at a lower cost and reducing environmental pollution generated by the use of chemical solvents in conventional extraction ways.

Due to these circumstances, ellagic acid production has been carried out by exploring fungal strains and fermentation processes using agro-industrial wastes. Buenrostro et al. [13] reported with ellagitannase as biocatalyst, which was attached to polyurethane foam, a continuous system for ellagic acid production SSF.

11.8   FACTORS INFLUENCING SOLID-STATE FERMENTATION (SSF)

11.8.1   MICROORGANISM

The most used microorganisms to perform in SSF are generally fungi, bacteria, and yeast, however, filamentous fungi are more suitable because they can tolerate lower water activity (WA) and can grow quickly on the substrate, besides, they help to get the products of interest because they produce a higher amount of enzymes [54].

11.8.2 SUBSTRATE

Agricultural and forest residues are the most commonly used substrate to perform a SSF due to great abundance and non-specific use, which represent a high source of contamination. Some examples of agriculture's residues reported in SSF processes are bagasse from sugarcane and cassava, as well as residues from some cereals, such as wheat, rice, and oats [55].

The husk of several fruits, coffee pulp, and other residues from different fruits and plants have also been used. All these substrates have a high content of molecules, such as, pectins, starch, hemicellulose, cellulose, lignin, etc., which is used by microorganisms' growth. The substrates usually contain all the sources necessary for the microorganism to grow. However, in several cases, it is necessary required to add some macro and micronutrients, such as phosphorus, sulfur, calcium, magnesium, iron, zinc, cobalt, and iodine, this is for optimal growth [55].

11.8.3 INOCULUM TYPE

The inoculum used is of importance to the efficiency of a SSF; usually, the medium is inoculated with spores instead of using vegetable cells, this because the spores carry out the same reactions as the fungus mycelium, besides, their isolation and the preparation of the inoculum is easier and can be store for later use [50]. However, some of the disadvantages they present are that their growth requires more time, the germination and growth conditions can vary according to the substrate and, finally, a higher amount of inoculum is needed.

11.8.4 MOISTURE AND WATER ACTIVITY (WA)

For a SSF to be carried out in the best way, it is important to take into account the humidity of the medium. Fungi are microorganisms that require an environment with adequate moisture for their growth. Low and high levels of moisture can cause an inappropriate absorption of nutrients. Therefore, affects the stability of enzymes and it hinders the growth of microorganisms [56].

SSF process uses humidity values between 30% and 85% in general, but it must generally be between 20% and 70% for fungal strains [57]. The water that the microorganisms require for their metabolism in a SSF is determined as the WA present in the substrate. WA can be defined as the available water that can react with the substrate. This largely depends on the ability of the substrate to bind water molecules.

The WA of the substrate decreases during the fermentation due to its dehydration or the accumulate of solutes in the substrate, and low WA, as well as moisture, affects growth, which leads to less biomass production [54]. WA and humidity are closely related so that if there a change in one of them, it will have a great impact on the other [58].

11.8.5   PH

The pH is another essential factor to consider in a SSF; changes in pH values according to the process carried out and its greatly influenced by the production of organic acids during fermentation, which causes decrease; on the contrary, the assimilation of organic acids presents and the alkalization of urea present can increase the pH.

Filamentous fungi are microorganisms that tolerate a wide pH range since they can see grow between values of 2 to 9, but their optimal pH is in the field of 3.8 to 6 [58].

The pH is an important factor to be considered during process fermentation, due to microorganisms' ability to grow at a low pH can significantly prevent contamination with other microorganisms. However, the control of pH in a SSF is hard, because the system is heterogeneous and because they are not equipment to measure the pH in stable systems. In order to control this situation, a buffer solution can be added, that does not interfere with the biological processes of the fermentation, or the substrate can be formulated from the beginning when considering the components and their ability to act as a buffer [59].

11.8.6   TEMPERATURE

The most important physical factor for the development of a SSF is the temperature: due to the enzymes and specific metabolites produced that are highly sensitive to temperature. The filamentous fungi stand temperatures range between 25 to 55°C, and the optimal temperature differs depending on the product of interest to be produced; therefore, they are the most efficient microorganisms to perform a SSF [60].

Temperature is a difficult factor to control since solid materials do not have adequate thermal conductivity; therefore, the devices normally used to conduct heat or energy do not work correctly on these substrates. Therefore, what is normally done to control the temperature to a certain extent is through the elimination of heat, mainly by aeration or evaporative cooling [61].

11.9   BIOTECHNOLOGICAL PRODUCTION OF ELLAGIC ACID

Organic solvents like methanol, acetone, ethyl acetate and hexane, are commonly used in conventional extraction techniques to recover phenolic compounds from plants. However, the environmental pollution toxicology is associated with the use of these solvents, that's why emergent technologies of extraction of bioactive compounds have improved the yield extraction and one of them is the biotechnological extraction. The use of bioprocesses such as enzymatic processes of extraction of bioactive compounds from agro-industrial products and fermentation have been reported by some authors [2].

There are several derivates of ellagic acid existing in plants, formed through metoxylation, glycosylation, and methylation of its hydroxyl groups [11]. Although under different reactions conditions, as microbial, enzymatic, or chemical conditions, ellagitannins change to free ellagitannin and their derivates changing reactivity, solubility, and mobility of these derivates and, a consequence of these reactions, its biological capacity depending on its structure.

Enzymes for ellagic acid production have been used but in SSF has been poorly evaluated, the action of one or more enzymes to ellagitannins biodegradation into ellagic acid have been reported by some research groups; due to this research Ascacio-Valdés et al. [2]

evaluated the activities of the enzymes on releasing ellagic acid from ellagitannins of pomegranate, reporting that only ellagitannase or ellagitannin acyl hydrolase was associated with the accumulation of ellagic acid in SSF.

Just as the use of enzymes in the production of ellagic acid, microbial hydrolysis has also been poorly evaluated in SSF. Ascacio-Valdés et al. [62] reported that in SSF, *A. niger GH1* was able to grow and also produce ellagitannase enzyme. The ester bonds among HHDP group and glycosides are degraded by ellagitannase which has a high specificity, allowing the ellagic acid accumulation in the model that the authors proposed.

The process of SSF consists that in the absence of water, the microbial growth and formation on surface and inside of a porous solid matrix [63], where simulating natural growth conditions, the substrate must contain enough moisture to allow the microorganism growth and metabolism [64]. Also, the critical limitation point is the water availability in SSF, which can interfere in the growth and metabolism of the microorganism; this availability water responds to WA which determines the equilibrium between the gaseous atmosphere with the substrate.

In research reported by Buenrostro et al. [13] the ellagitannase production of pomegranate was performed in several sterile columns considered bioreactors packed with a homogeneous mixture containing corn cobs, coconut husk, sugarcane bagasse and candelilla stalks. The results of this research settle that all the tested agro-industrial products have a good potential to be used as support in SSF process.

In Table 11.5, can be compared the values obtained of ellagic acid of some research with SSF extraction method.

TABLE 11.5   Ellagic Acid Obtained with a Solid-State Fermentation Extraction Method

Substrate	Portion	References
Pomegranate peel	21 mg/g	[65]
Pomegranate	42.02 mg/g	[2]
Inert material (PUF)	231.22 mg/g	[13]
Tea waste	42.35 µg/g	[66]
Pomegranate peel	132 mg/g	[15]

In Table 11.6 can be compared the values of ellagic acid with different methods of extraction, among that in some studies (like conventional with the use of harmful solvents), the value of ellagic acid is higher than biotechnological extraction method, however, must be remembered that the disadvantages of conventional methods are that the use of them is harmful to the environment.

TABLE 11.6   Ellagic Acid Values of Different Extraction Methods

Substrate	Extraction Method	Portion	References
Strawberries	Sulfuric acid/pyridine	226 ± 5.40 mg/100 g	[67]
Blackberries	Sulfuric acid/pyridine	82.71 ± 6.56 mg/100 g	[67]
Pomegranate peel	Solid-state fermentation	21 mg/g	[65]
Pomegranate	Solid-state fermentation	42.02 mg/g	[2]
Inert material (PUF)	Solid-state fermentation	231.22 mg/g	[13]

Compared with chemical methods in ellagic acid production, this research shows that the development of a bioprocess for ellagitannase production could offer economic and environmental advantages.

11.10   CONCLUSION

The ellagitannins biotransformation results in a bioactive compound of a high biological relevance such as the ellagic acid; however, this is a process that has been poorly explored. Several studies have been realized about the extraction of this compound by conventional technologies or even by emergent technologies such a microwave. However, it has been reported that traditional extraction methods have operational disadvantages as for recovery performances of these phenolic compounds, the environmental pollution, and the use of dangerous solvents.

So, with this explanation, the SSF of ellagic acid has come up as an alternative for this bioactive compound extraction. In rambutan peel, in particular, the current information regarding using bioprocesses for the extraction of compounds is scarce compared to the reference information used by emergent technologies as US.

It is necessary to evaluate new extraction sources of these phenolic compounds and used them to provide the development of new scientific knowledge using the natural resources of our country; to obtain unique and different results, it must search new ways to obtain ellagitannins for a bigger picture in the studies that in a future could be compared, confirmed, and get to get a discussion with the current information.

ACKNOWLEDGMENTS

Autor Cerda-Cejudo N.D. thanks the Mexican Council for Science and Technology (CONACYT) for the financial support for the development of the project (Master) in Food Science and Technology offered by the Autonomous University of Coahuila.

KEYWORDS

- *Aspergillus niger*
- **bioactive**
- **dehydrodigalloyl**
- **ellagic acid**
- **hexahydroxydiphenoyl**
- **polyphenols**

REFERENCES

1. Quideau, S., Deffieux, D., Douat-Casassus, C., & Pouységu, L., (2011). Plant polyphenols: Chemical properties, biological activities, and synthesis. *Angew. Chem. Int. Ed., 50*, 586–621.
2. Ascacio-Valdés, J. A., Buenrostro, J. J., De La Cruz, R., Sepúlveda, L., Aguilera, A. F., Prado, A., Contreras, J. C., Rodríguez, R., & Aguilar, C. N., (2014). Fungal biodegradation of pomegranate ellagitannins. *J. Basic Microbiol., 54*, 28–34.
3. Barbeheen, R. V., & Constabel, C. P., (2011). Tannins in plant-herbivore interactions. *Phytochem., 72*, 1551–1565.
4. Jourdes, M., Pouysegu, L., Deffieux, D., Teissedre, P. L., & Quideau, S., (2013). Hydrolizable tannins: Gallotannins and ellagitannins. In: Ramawat, K. G., & Mérillon, J. M., (eds.), *Natural Products: Phytochemistry Botany and Metabolism of Akaloids Phenolics and Terpenes* (pp. 1975–2010). Springer.

5. Aguilar-Zarate, P., Wong-Paz, J. E., Buenrostro-Figueroa, J. J., Ascacio, J. A., Contreras-Esquivel, J. C., & Aguilar, C. N., (2017). Ellagitannins: Bioavailability, purification and biotechnological degradation. *Mini Rev. Med. Chem., 18*, 1244–1252.

6. Pouységu, L., Deffieux, D., Malik, G., Natangelo, A., & Quideau, S., (2011). Synthesis of ellagitannin natural products. *J. Nat. Prod. Rep., 28*, 853–874.

7. Robledo, A., Aguilera-Carbó, A., Rodríguez, R., Martínez, J. L., Garza, Y., & Aguilar, C. N., (2008). Ellagic acid production by *Aspergillus niger* in solid state fermentation of pomegranate residues. *J. Ind. Microbiol. Biotechnol., 35*, 507–513.

8. Cheynier, V., Comte, G., Davies, K. M., Lattanzio, V., & Martens, S., (2013). Plant phenolics: Recent advances on their biosynthesis, genetics, and ecophysiology. *Plant Physiol. Biochem., 72*, 1–20.

9. Chen, Y., & Hagerman, A. E., (2004). Characterization of soluble non-covalent complexes between bovine serum albumin and beta-1, 2, 3, 4, 6-penta-O-galloyl-D-glucopyranose by MALDI-TOF MS. *J. Agric. Food Chem., 52*, 4008–4011.

10. Chávez, M. L., Guyot, S., Rodríguez, R., Prado, A., & Aguilar, C. N., (2014). Production profiles of phenolics from fungal tannic acid biodegradation in submerged and solid-state fermentation. *Process Biochem., 49*, 541–546.

11. Landete, J. M., (2011). Ellagitannins, ellagic acid and their derived metabolites: A review about source, metabolism, functions and health. *J. Food Res. Inter., 44*, 1150–1160.

12. Dhingra, D., & Jangra, A., (2014). Antiepileptic activity of ellagic acid, a naturally occurring polyphenolic compound, in mice. *J. Func. Foods, 10*, 364–369.

13. Buenrostro-Figueroa, J., Huerta-Ochoa, S., Prado-Barragán, A., Ascacio-Valdés, J., Sepúlveda, L., Rodríguez, R., Aguilera-Carbó, A., & Aguilar, C. N., (2014). Continuous production of ellagic acid in a packed-bed reactor. *Process Biochem., 49*, 1595–1600.

14. Sepúlveda, L., Wong-Paz, J. E., Buenrostro-Figueroa, J., Ascacio-Valdés, J. A., Aguilera-Carbó, A., & Aguilar, C. N., (2018). Solid state fermentation of pomegranate husk: Recovery of ellagic acid by SEC and identification of ellagitannins by HPLC/ESI/MS. *Food Biosci., 22*, 99–104.

15. Sepúlveda, L., Aguilera-Carbó, A., Ascacio-Valdes, J., Rodriguez-Herrera, R., Martínez, J. L., & Aguilar, C. N., (2012). Optimization of ellagic acid accumulation by *Aspergillus niger* GH1 in solid state culture using pomegranate shell powder as a support. *Process Biochem., 47*, 2199–2203.

16. Clifford, M. N., & Scalbert, A., (2000). Ellagitannins-nature, occurrence and dietary burden. *J. Sci. Food Agric., 80*, 1118–1125.

17. Wang, Y., Zhang, H., Liang, H., & Yuan, Q., (2013). Purification, antioxidant activity and protein-precipitating capacity of punicalin from pomegranate husk. *Food Chem., 138*, 437–443.

18. Lu, J., Ding, K., & Yuan, Q., (2008). Determination of punicalagin isomers in pomegranate husk. *Chromatographia, 68*, 303–306.

19. Park, M., Cho, H., Jung, H., Lee, H., & Hwang, K. T., (2014). Antioxidant and anti-inflammatory activities of tannin fraction of the extract from black raspberry seeds compared to grape seeds. *J. Food Biochem., 38*, 259–270.

20. Versari, A., Du Toit, W., & Parpinello, G. P., (2013). Oenological tannins: A review. *Aust. J. Grape Wine Res., 19*, 1–10.

21. Thitilertdecha, N., Teerawutgulrag, A., Kilburn, J. D., & Rakariyatham, N., (2010). Identification of major phenolic compounds from *Nephelium lappaceum* L. and their antioxidant activities. *Molecules, 15,*1453–1465.

22. Hager, T. J., Howard, L. R., Liyanage, R., Lay, J. O., & Prior, R. L., (2008). Ellagitannin composition of blackberry as determined by HPLC-ESI-MS and MALDI-TOF-MS. *J. Agric. Food Chem., 3,* 661–669.

23. Fukuda, T., Ito, H., & Yoshida, T., (2004). Effect of the walnut polyphenol fraction on oxidative stress in type 2 diabetes mice. *Biofactors., 21,* 251–253.

24. Rohman, A., (2017). Physico-chemical properties and biological activities of rambutan (*Nephelium lappaceum L.)* fruit. *Res. Journal Phytochemistry, 11,* 66–73.

25. Akhtar, M. T., Ismail, S. N., & Shaari, K., (2017). Rambutan (*Nephelium lappaceum* L.). In: Yahia, E. M., (ed.), *Fruit and Vegetable Phytochemicals* (pp. 1227–1234). John Wiley & Sons Ltd: México.

26. Suganthi, A., & Marry, J. R., (2016). *Nephelium lappaceum* (L.): An overview. *Inter. J. Pharm. Sci. Res., 1,* 36–39.

27. Mahmood, K., Fazilah, A., Yang, T. A., Sulaiman, S., & Kamilah, H., (2018a). Valorization of rambutan (*Nephelium lappaceum*) by-products: Food and non-food perspectives. *Inter. Food Res. J., 25,* 890–902.

28. Morton, J., (1987). *Rambutan: Fruits of Warm Climates.* Miami FL.

29. Mahmood, K., Kamilah, H., Alias, A. K., & Ariffin, F., (2018b). Nutritional and therapeutic potentials of rambutan fruit (*Nephelium lappaceum L.*) and the by-products: A review. *J. Food Meas. Charact., 12,* 1556–1571.

30. Santana-Meridas, O., González-Coloma, A., & Sánchez-Vioque, R., (2012). Agricultural residues as a source of bioactive natural products. *Phytochem. Rev., 11,* 447.

31. Solís-Fuentes, J. A., Camey-Ortiz, G., Hernández-Medel, M. R., Pérez-Mendoza, F., & Durán-De, B. C., (2010). Composition, phase behavior and thermal stability of natural edible fat from rambutan (*Nephelium lappaceum* L.) seed. *Bioresour. Technol., 101,* 799–803.

32. Serida, N. H., Nazaruddin, R., Nazanin, V., & Mamot, S., (2012). Physicochemical and nutritional composition of rambutan Anak Sekolah (*Nephelium lappaceum L.*) seed and seed oil. *Pakistan J Nutr., 11,* 1073–1077.

33. Fraire, V. G., (2001). *El rambután: Alternativa para la producción frutícola del trópico húmedo de México* (1st edn.). SAGARPA: INIFAP, Centro de Investigación Regional del Pacífico Sur, Campo Experimental Rosario Izapa, Fundación Produce: Tuxtla Chico, Chiapas.

34. Fila, W., Itam, E., & Johnson, J., (2013). Comparative proximate compositions of watermelon *Citrullus lanatus*, squash *Cucurbita pepo'l* and rambutan *Nephelium lappaceum*. *Inter. J. Sci. Technol., 2,* 81–88.

35. Issara, U., Zzaman, W., & Yang, T. A., (2014). Rambutan seed fat as a potential source of cocoa butter substitute in confectionary product. *Inter. Food Res. J., 21,* 25–31.

36. Vargas, A., (2003). Descripcion morfologica y nutricial del fruto de rambutan (*Nephelium lappaceum*). *Agronomía Mesoamericana, 14,* 201–206.

37. Hernández, C., Ascacio-Valdes, J., De La Garza, H., Wong-Paz, J., Aguilar, C. N., & MartinezAvila, G. C., (2017). Polyphenolic content, *in vitro* antioxidant activity and

chemical composition of extract from *Nephelium lappaceum L.* (Mexican rambutan) husk. *Asian Pac. J. Trop. Med., 10*, 1201–1205.

38. Oliveira, E. I. S., Santos, J. B., Paula, A., Goncalves, B., & Jose, N. M., (2016). Characterization of the rambutan peel fiber (*Nephelium lappaceum*) as a lignocellulosic material for technological applications. *Chem. Eng. Trans., 50*, 391–396.

39. Okonogi, S., Duangrat, C., Anuchpreeda, S., Tachakittirungrod, S., & Chowwanapoonpohn, S., (2007). Comparison of antioxidant capacities and cytotoxicities of certain fruit peels. *Food Chem., 103*, 839–846.

40. Palanisamy, U. D., Cheng, H. M., Masilamani, T., Subramaniam, T., Ling, L. T., & Radhakrishnan, A. K., (2008). Rind of the rambutan, *Nephelium lappaceum*, a potential source of natural antioxidants. *Food Chem., 109*, 54–63.

41. Hernández-Hernández, C., Aguilar, C. N., Rodriguez-Herrera, R., Flores-Gallegos, A. C., Morlett-Chávez, J., Govea-Salas, M., & Ascacio-Valdés, J. A., (2019). Rambutan (*Nephelium lappaceum*L.): Nutritional and functional properties. *Trends Food Sci. Technol., 85*, 201–210.

42. Škerget, M., Kotnik, P., Hadolin, M., Hraš, A. R., Simonič, M., & Knez, Ž., (2005). Phenols, proanthocyanidins, flavones and flavonols in some plant materials and their antioxidant activities. *Food Chem., 89*, 191–198.

43. Kronholm, J., Hartonen, K., & Riekkola, M. L., (2007). Analytical extractions with water at elevated temperatures and pressures. *TRAC Trends Anal. Chem., 26*, 396–412.

44. Ahmad, A., Palanisamy, S. A., Tejo, U. D., Chew, B. A., Tham, M. F., & Hassan, S., (2017). Geraniin extracted from the rind of *Nephelium lappaceum* binds to dengue virus type-2 envelope protein and inhibits early stage of virus replication. *Virol. J., 14*, 1–13.

45. Domínguez-Rodríguez, G., Marina, M. L., & Plaza, M., (2017). Strategies for the extraction and analysis of non-extractable polyphenols from plants. *J. Chrom. A*, 1–15.

46. Plaza, M., & Rodríguez-Meizoso, I., (2013). Advanced extraction processes to obtain bioactives from marine foods. In: Hernández-Ledesma, B., & Herrero, M., (eds.), *Bioactive Compounds from Marine Foods* (pp. 343–371). John Wiley & Sons Ltd: Chichester, UK.

47. Zhang, S., Li, L., Cui, Y., Luo, L., Li, Y., Zhou, P., & Sun, B., (2017). Preparative high-speed counter-current chromatography separation of grape seed proanthocyanidins according to degree of polymerization. *Food Chem., 219*, 399–407.

48. Herrero, M., Castro-Puyana, M., Mendiola, J. A., & Ibañez, E., (2013). Compressed fluids for the extraction of bioactive compounds. *TRAC Trends Anal. Chem., 43*, 67–83.

49. Cannel, E., & Moo-Young, M., (1980). Solid-state fermentation systems. *Process. Biochem., 4*, 2–7.

50. Krishna, C., (2005). Solid-state fermentation systems—An overview. *Crit. Rev. Biotechnol., 25*, 1–30.

51. Torres-León, C., Ramírez-Guzmán, N., Ascacio-Valdés, J., Serna-Cock, L., Dos, S. C. M. T., Contreras, E. J. C., & Aguilar, C. N., (2019). Solid-state fermentation with *Aspergillus niger* to enhance the phenolic contents and antioxidative activity

of Mexican mango seed: A promising source of natural antioxidants. *LWT Food Sci. Technol., 112*, 108–236.

52. Buenrostro-Figueroa, J., Ascacio-Valdés, J. A., Sepúlveda, L., Prado-Barragán, A., Aguilar-González, M. A., & Aguilar, C. N., (2018). Ellagic acid production by solid-state fermentation influenced by the inert solid supports. *Emir. J. Food Agric., 30*, 750–757.

53. Sun, L., Zhang, H., & Zhuang, Y., (2012). Preparation of free, soluble conjugate, and insoluble-bound phenolic compounds from peels of rambutan (*Nephelium lappaceum*) and evaluation of antioxidant activities *in vitro. J. Food Sc., 77*, 1750–3841.

54. Raimbault, M., (1998). General and microbiological aspects of solid substrate fermentation. *Elect. J. Biotechnology, 1*, 1–15.

55. Soccol, C. R., Ferreira Da, C. E. S., Letti, L. A. J., Karp, S. G., Woiciechowski, A. L., & Vandenberghe, L. P. S., (2017). Recent developments and innovations in solid state fermentation. *Biotechnol. Res. Innov., 245*, 1727–1739.

56. Moo-Young, M., Moriera, A. R., & Tengerdy, R. P., (1983). Principles of solid state fermentation. In: Smith, J. E., Berry, D. R., & Kristiansen, B., (eds.), *The Filamentous Fungi* (Vol. 4, pp. 117–144). Edward Arnold Publishers: London.

57. Pandey, A., Soccol, C. R., Rodríguez-León, J. A., & Nigam, P., (2001). *Solid-State Fermentation in Biotechnology - Fundamentals and Applications* (pp. 100–221.). Asiatech Publishers: New Delhi.

58. Gowthaman, M. K., Krishna, C., & Moo-Young, M., (2001). Fungal solid state fermentation - an overview. In: Khachatourians, G., & Arora, D. K., (eds.), *Applied Mycology and Biotechnology* (pp. 305–352). Elsevier.

59. Lonsane, B. K., Saucedo-Castañeda, S., Raimbault, M., Roussos, S., Viniegra-González, G., Ghildyal, N. P., Ramakrishna, M., & Krishnaiah, M. M., (1992). Scale-up strategies for solid state fermentation systems. *Proc. Biochem., 27*, 259–273.

60. Yadav, J. S., (1988). SSF of wheat straw with alcaliphilic *Coprinus. Biotechnol. Bioeng., 31*, 414–417.

61. Trilli, A., (1986). Scale up of fermentation. In: Demain, A. L., & Solomon, N. A., (eds.), *Industrial Microbiology and Biotechnology* (pp. 227–307). American Society of Microbiology: Washington, U.S.A.

62. Ascacio-Valdés, J. A., Buenrostro, J. J., De La Cruz, R., Sepúlveda, L., Aguilera, A. F., Prado, A., Contreras, J. C., et al., (2013). Fungal biodegradation of pomegranate ellagitannins. *J. Basic Microbiol., 54*, 28–34.

63. Barrios-González, J., (2012). Solid-state fermentation: Physiology of solid medium, its molecular basis and applications. *Process Biochem., 47*, 175–185.

64. Orzúa, M. C., Mussatto, S. I., Contreras-Esquivel, J. C., Rodríguez, R., De La Garza, H., Teixeira, J. A., & Aguilar, C. N., (2009). Exploitation of agro industrial wastes as immobilization carrier for solid-state fermentation. *Ind. Crops Prod., 30*, 24–27.

65. Sepúlveda, L., De La Cruz, R., Buenrostro, J. J., Ascacio-Valdés, J. A., Aguilera-Carbó, A. F., Prado, A., & Aguilar, C. N., (2016). Effect of different polyphenol sources on the efficiency of ellagic acid release by *Aspergillus niger. Rev. Arg. Microbiol., 48*, 71–77.

66. Paranthaman, R., Kumaravel, S., & Singaravidel, K., (2013). Development of bioprocess technology for the production of bioactive compound, ellagic acid from tea waste. *Curr. Res. Microbiol. Biotechnol., 1*, 270–273.

67. Márquez-López, A., Ayala-Flores, F., Macías-Pureco, S., Chávez-Parga, M. C., Valencia-Flores, D. C., Maya-Yescas, R., & González-Hernandéz, J. C., (2019). Extract of ellagitannins starting with strawberries (*Fragaria sp.*) and blackberries (*Rubus sp.*). *Food Sci. Technol., 40*, 430–439.

68. Sekar, M., Jaffar, F. N. A., Zahari, N. H., Mokhtar, N., Zulkifli, N. A., Kamaruzaman, R. A., & Abdullah, S. (2014). Comparative evaluation of antimicrobial properties of red and yellow rambutan fruit peel extracts. *Annu. Rev. Plant. Biol., 4,* 3869.

69. Thitilertdecha, N., Teerawutgulrag, A., & Rakariyatham, N. (2008). Antioxidant and antibacterial activities of *Nephelium lappaceum* L. extracts. *LWT – Food Sci. Technol., 41,* 2029–2035.

70. Khaizil Emylia, Z., Nik Aina, S. N. Z., & Mohd Dasuki, S. (2013). Preliminary study on anti-proliferative activity of methanolic extract of *Nephelium lappaceum* peels towards breast (MDA-MB-231), cervical (HeLa) and osteosarcoma (MG-63) cancer cell lines. *J. Environ. Public. Health., 4,* 66–79.

71. Morshed, T. M. I., Dash, P. R., Ripa, F. A., Foyzun, T., & Mohd, A. S. (2014). Evaluation of pharmacological activities of methanolic extract of *Nephelium lappaceum* L. seeds. *Int. J. Pharmacogn. Phytochem. Res., 1,* 632–639.

72. Domínguez-Rodríguez, G., Marina, M. L., & Plaza, M. (2017). Strategies for the extraction and analysis of non-extractable polyphenols from plants. *J. Chromatogr. A., 1514,* 1–15.

CHAPTER 12

Novel Methods of Food Preservation

C. GUILLERMO VALDIVIA-NÁJAR and LORENA MORENO-VILET

CONACYT-Department of Food Technology, Centre of Research and Assistance in Technology and Design of the State of Jalisco, A.C. Zapopan, Jalisco, México, E-mail: gvaldivia@ciatej.mx (C. G. Valdivia-Nájar)

ABSTRACT

The most crucial challenge to preserving foodstuffs is microbiological stability since it determines the food safety and shelf life of products. Since ancient times, many methods have been used to stop or slow down food spoilage and consequently minimize any chance of foodborne illness, and these methods have evolved over time. This evolution has accelerated a lot in recent years due to consumer demands, which depend on changes in lifestyle, consumption habits, chronic diseases in the modern population, and the green and efficient methods demanded by the industry. Today consumers demand high nutritional value, functional properties, health benefits and fresh-like appearance, in addition to the strictly necessary food safety; thus, new technologies seek to meet these needs. In this chapter, the main conventional physical and chemical methods are discussed to introduce the novel thermal and non-thermal technologies used in food preservation, such a microwave, radiofrequency, infrared, ohmic heating, high-pressure processing, ultrasound, pulsed electric fields, pulsed light & UV, ionizing radiation and cold plasma.

12.1 INTRODUCTION

The main objective of the food preservation is to get innocuous products with a long shelf-life, it means to handle food in such a way as to stop or slow down spoilage and consequently minimize any chance of foodborne illness and in turn maintaining the optimum nutritional value, flavor, and texture. Microbiological stability is the main limitation of quality and determines the shelf life of fresh cut products. However, ripening, and food processing also lead to cell damage, changes in the physical structures and the release of nutrients and thus causing the optimal conditions for the development of microorganisms and physical, chemical, biological, and biochemical changes that alter the quality of food. However, the "food preservation" concept has gradually changed throughout years. Nowadays, the term preservation is not only associated to methods to ensure the food safety but also characteristics such as nutritive, functional, health-benefits, and fresh-like appearance are coveted. This modification of the concept has been largely influenced by changes in the lifestyle, consumption habits and presence of diseases in the modern population as well as the industry demand for green and efficient methods. Therefore, the initial and conventional methods used to ensure the food innocuity begin to be revaluated, replaced, or even fashioned to novel technologies with the main objective of produce safe, healthy, and attractive food products. So, in this chapter the principal conventional and emerging technologies used in food preservation will be discussed.

12.2   CONVENTIONAL METHODS

Based on the kind of processing, the conventional method for food preservation can be classified such as chemicals and physicals. Into the chemical processes, the application of chemical substances to perform washing, dipping or conservative additives, while physical methods imply the use of thermal and cooling methods and reduction of water activity (WA).

12.3   CHEMICAL METHODS

Among conventional methods, the use of chemical compounds or additives to perform conservation treatments such as washing, immersion (dipping) and gaseous stand out. These chemical substances can be found in nature or synthesized in a laboratory and can be classified such as antimicrobials, antioxidants, or root inhibitors. So, chemical compounds can be applied in a wide number of methods such as washing, dipping, gas, etc. Usually, the washing treatments are used not only to remove residues and possible contaminants added during harvest but also to eliminate microbes and tissue fluids released during the food processing. Moreover, washing with antimicrobial and antioxidant agents are commonly used to improve the quality of the fresh-cut products. Hypochlorite and potassium meta bisulfate are the most used chemicals to maintain the fresh-like quality of fresh products [1]. In addition, washing treatments with antimicrobial agents such as can be used to reduce microbial counts and maintain physicochemical characteristics during storage of fresh-cut products [2].

Similar to washing treatments, the dipping method is another popular technique used to preserve fresh commodities, reducing the microbial growth, and avoiding the physiochemical degradation of those foods. Hydrogen peroxide immersion treatments have shown a positive effect in the reduction of microbial counts; however, it has a negative effect on the antioxidant capacity and the content of bioactive compounds in fresh cut tomatoes [3]. Some studies propose the use of immersion washes with agents such as citric acid, chlorine dioxide, ozone, and peroxy-acetic acid to induce a reduction of the microbial loads. The application of edible films is also effective in the conservation of FFCs [4]. The ability of these treatments to maintain the quality of FFCs is attributed to the formation of a semipermeable film in the product, altering the permeability of water, oxygen, and carbon dioxide, reducing respiration, and delaying the senescence of the fruit [5]. Also, the chemical volatile compounds have shown excellent results in the preservation of food quality, reducing the metabolic respiration, and thus delaying the chemical changes involved in food ripening. The use of 1-methyl-cyclopropane (1-MCP) can retain the physicochemical characteristics of fresh tomato cut during storage at 5°C [6]. The use of volatile compounds such as jasmonate and ethanol reduces the microbial load and maintains the content of bioactive compounds during the refrigerated storage (5°C) of the cut products [7].

12.4 PHYSICAL METHODS

12.4.1 THERMAL PROCESSING

The thermal processing methods have been considered the most common way of preserving food for more than 150 years, since the great majority of pathogenic microorganisms are killed at high temperatures. In those processes, the food is placed into a unit where is heated to a certain temperature for a certain time and thus, the microorganism and/or enzymatic inactivation is promoted. The principal conventional thermal methods used to ensure innocuity and preserve the food commodities are pasteurization, sterilization, hot air drying and ultrahigh temperature which are applied to induce changes that improve food quality by different food production operations. Both, temperature, and time, are principal study parameters influencing the effectiveness of thermal methods. Therefore, design of thermal processing is based on the inactivation of the most thermal-resistant undesirable microorganism or enzyme in the food product, and thus ensuring the safety and quality of the processed food.

The main thermal processes used for food preservation are commonly described as commercial sterilization and pasteurization. The objective of the commercial sterilization is to ensure the food stability under ambient conditions from months up to years and where no spores and vegetative cells that can grow under that storage conditions. Thereby, the final shelf-life of the sterilized food products are not only determined by the microbial presence but also by the physicochemical and/or sensorial characteristics. On the other hand, pasteurization is a milder thermal process, where pathogenic microorganisms, notably vegetative cells, yeasts, and mold are inactivated. However, to guarantee the food stability and safety, pasteurization needs to be used with a complementary physical method or chemical preservatives. Pasteurization is carried out at temperatures between 65° and 100°C for a specific time and is usually applied in a wide variety of beverages. Pasteurized food products have a shelf life from a few days (e.g., pasteurized milk and product with high aw) to months or years (e.g., juices, and products with low aw).

12.4.2   COOLING-FREEZING

The temperature is a physical measure that relates the internal energy of the system, which is directly proportional to their level of molecular agitation and vibration. At lower temperatures, the energy level is smaller, as will be slower the reactions involving them [8]. The above principle is applied to food preservation by cooling at temperatures between 4 and 8°C, where the water in food products keep in liquid state and the physicochemical, microbiological, and biochemical reactions responsible for food spoilage occur but very slowly. The cooling/freezing process is used in the food industry for ensuring the shelf-life extension of products with minimal changes in sensory and nutritional attributes. However, it is important to remark that this process does not result in microbial inactivation, because cooling only reduce the microbial growth rate, depending of the nature of the microorganism either thermophiles, mesophilic, psychrophilic, or psychotropic, since each microorganism has a minimum temperature of development, and below this temperature it remains in a latent state.

Freezing is a more extreme process where the temperature reach values below 0°C for long-term preservation of food products. During freezing, the latent heat is removed by lowering the temperature of the food and boosting the ice crystals formation, reducing the WA and therefore the growth of microorganisms and enzymatic reactions [9]. In fact, chilled or refrigerated foods have a medium shelf-life of days or weeks, while the shelf-life of frozen foods can achieve long shelf-life periods of several months.

In regard to safety in frozen products, it supposed that microorganisms enter into latency state, and in some cases inactivated due to the low temperatures, however, in cases when the reduction of microbial load is required, additional method of preservation as pasteurization should be used for safety assurance in consumption. It is considered that frozen foods are preferred by consumers since it ensure better organoleptic properties, nutritional value, and above all freshness than dried products, concentration, and pasteurization. Within the main applications of this method is for preserve most fresh vegetables, fruits, meats, fish, breads, and cakes.

12.4.3   WATER ACTIVITY (WA) REDUCTION

The water activity (a_w) concept has been very useful in food preservation since it has served as the basis for the design of different preservation processes and the creation of new food products. WA is a measure of the amount of water available in the food and it is defined as the ratio between the vapor pressure of water exerted by the food and the vapor pressure of pure water at the same temperature (FAO: *Food and Agriculture Organization*). This parameter varies between 0 and 1, the higher the value of a_w the food is more unstable by spoilage reactions. There are critical

a_w values that correspond to maximum and minimum values for both physical and chemical reactions, values between 0.2 and 0.3 represents the region of optimum moisture content in which dehydrated food products have the maximum shelf life. Above this value, chemical reactions that need an aqueous phase begin, and at lower values, the rate of lipid oxidation increases, compromising the shelf life of food products. At values of a_w between 0.35 and 0.45 physical changes such as loss of crunchiness, adhesiveness of powdered products and hardening of sweets begin. Within the methods focus on a_w reduction are concentration and drying processes. In the concentration process, the amount of water removed from the food is relatively small. As a result, the final a_w still allows microbial growth and an additional preservation method is needed as thermal pasteurization processes. The osmotic dehydration is the most widely used process for the concentration of solids, while the concentration by boiling, use of membrane technology and vacuum concentration are the most used for liquid products [8]. As a matter of fact, drying is one of the oldest methods of preserving food and can be specified as a simultaneous heat and mass transfer operation in which the WA of a foodstuff is decreased by the removal of water by evaporation into an unsaturated gas stream. The resulting a_w, therefore, makes the product stable, allowing its conservation for a long period. Drying using hot air in convective ovens or spray drying for powders and freeze-drying are the commonly used methods for drying food.

12.5   INNOVATIVE AND EMERGING METHODS

Until recently food commodities were processed by conventional chemical (washing, dipping, additives) and physical techniques (thermal, drying, freezing, and cooling) to increase shelf life. However, not only the degradation and loss of the sensory and physiochemical attributes such as color, flavor, and texture but also the nutrients and bioactive compounds contents in food are highly affected by those treatments.

The growing consumers interest for fresh-like, safe, and convenient food products have prompted to the researchers to innovate and develop processing methods capable to preserve the nutritional compounds for longer periods. In this context, non-conventional, energy efficient and innovative processes to ensure the safety, quality, and maintenance of health-related compounds into food matrixes have been studied and evaluated during last years. Those emerging technologies, have been classified into two groups: (1) thermal such as microwaves (MW), radiofrequency (RF), infrared (IR) and ohmic heating (OH); and (2) nonthermal such as high-pressure processing (HPP), ultrasounds (US), pulsed electric fields (PEF), pulsed light (PL) and UV processing (PL and UV), ionizing radiation (IOR) and cold plasma (CP). Emerging thermal technologies has exhibit great results on the microbial inactivation while the nonthermal methods have shown better quality parameters in the products such a higher content of bioactive compounds and better sensory attributes in different food systems. However, the effectiveness of those processes to inactivate microorganism or enzymes and to maintain the quality of food products is conditioned and depends on intrinsic and extrinsic factors. The action mechanisms and effectiveness of the innovative and emerging methods are discussed in this section.

12.5.1 THERMAL METHODS

Based on the premise that the high temperature short time (HTST) process is ideal for producing safe food while maintaining food quality. The novel thermal processes such as RF, MW, IR and OH, focus on achieving a fast heating, either by volumetric or electrical heating. Compared to the conventional process, where heat is transferred slowly by convection, conduction, or radiation from outside to inside the food, causing it to overheat, the novel thermal technologies achieve the desirable process temperature in less time. Microwave (MW) and RF heating are also called dielectric heating and provide volumetric heating, where the heat is generated inside the material via absorption of electromagnetic energy from the applied field. In the case of OH technology, the electrical resistance of foods is used to directly convert the electricity into heat. Both, dielectric, and OH are direct methods in which heat is generated within the product, whereas IR heating is an indirect method and the energy is applied to the surface of a food by radiation and then, converted to heat [10]. The efficacy and mechanisms of action of thermal methods are discussed along this section.

12.5.2   *MICROWAVE HEATING (MW)*

Microwave heating is also known as dielectric heating and has been the most popular emerging technology in food processing for domestic and industrial uses due to its advantages of high heating rates, short processing time, ease of operation, and lower maintenance requirements [11]. Jiao et al. [12] stated that MW are electromagnetic waves propagate in free space at the speed of light, with a frequency range from 300 MHz to 300 GHz but for industrial, scientific, and medical (ISM) applications only the frequencies of 915 MHz and 2,450 MHz are permitted. Moreover, dipole rotation and ionic conductivity are predominant mechanisms for dielectric heating, the first one as a cause of friction among surrounding water molecules, while ionic conduction as a corresponding motion of dissociative ions of food as salts. Both mechanisms are dependent on temperature and frequency of the field as well as the moisture content and salt concentration in foodstuff.

Microwave heating is usually applied for drying, pasteurization, sterilization, thawing, tempering, precooking, baking of food materials, etc., and other food production operations. However, its main use is still a domestic application for the reheating of prepared foods at home. The mechanism of microorganism inactivation is by thermal effect, although other additional non-thermal effects have been reported, they have not been adequately clarified. From the point of view of food safety, the main problem of MW refers to the poor distribution of heat or non-uniform temperature, especially in solid products, adversely affecting both food quality and safety [11, 13]. This non-uniform heating pattern is also due to the low penetration depth of MW, so the effectiveness of the process depends on the size, shape, and nature (liquids/solids) of the food. For liquid food MW heating has demonstrated great ability for the destruction and/or inactivation of pathogenic/spoilage microorganisms and enzymes, thus, the fresh like characteristics of products are preserved, especially in a continuous fluid system [13]. For solid and semisolid foods, MW also is an effective means for developing short-time in-food package sterilization and pasteurization processes [14]. Due to higher heating rate and consequently shorter processing time, MW pasteurized products have better sensory and nutritional quality compared to those conventional pasteurized products.

12.5.3   RADIOFREQUENCY (RF)

Radiofrequency (RF) heating is a technology quite similar to MW technology and has been used since 1940 to cook various processed meat products, to heat bread, to dehydrate and blanch vegetables, etc. [15]. The frequency range of RF is lower but a longer wavelength than MWs and cover from 3 KHz to 300, however for ISM applications only the frequencies of 13.56, 27.12 and 40.68 MHz are permitted. The wavelength at RF frequencies ranges from 22 to 360 times as long as that of the 2 commonly used in microwave frequencies, which allows RF energy to penetrate dielectric materials as food products, more deeply than MW energy, thus a uniform electric field distribution results, for this reason RF is of greater industrial interest [12]. Despite these advantages over MW, currently the application of RF for food heating is used less frequently than MW heating. Besides, RF heating, also has been proposed for the treatment of seeds and product disinfestation or disinfection like bulk material with relatively larger dimensions [16]. Some researchers have been studied this technology to produce ready-to-eat packaged food products either pasteurized or sterilized such as egg and shellfish that are heat sensitive products with high nutritional value. Within the main problems implicated in this technology are the estimated high operational cost of using RF and technical problems, such as dielectric breakdown and thermal runaway heating, which can be damaging both the product and the package [17]. Added to the previous ones, there is also, the non-uniformity in heat distribution that could affect the food safety, thus specific studies for each product and presentation are needed to produce safe and shelf-stable foods.

12.5.4   INFRARED HEATING (IR)

IR heating is a third technology along with MW and RF based on the electromagnetic waves, but this technology operates at frequencies higher than 300 GHz or wavelength range from 0.7 to 1,000 mm. IR radiation may be divided into three general regions: near IR radiation (NIR) ranging from 0.7 to 1.4 mm, medium IR radiation (MIR) ranging from 1.4 to 3 mm, and far IR radiation (FIR) ranging from 3 to 1,000 mm [18]. Most of the food components absorb radiative energy at FIR region, so it is used for food processing. The main advantages of IR technology are uniform heating, high heat transfer rate, reduced processing time and low energy consumption, as well as increasing quality and safety of the product, with a versatile, simple, and compact equipment [19]. IR technology was first industrially used in the 1930s, however the development of applications in food processing and agricultural products has gone slowly since then. Within the international applications of IRH for food processing are blanching, drying, peeling, baking, and roasting. Apart from this application of surface heating of foods, IR radiation could conveniently be used for decontamination and disinfection of food and food-contact surfaces since its effectiveness against pathogenic bacteria has been demonstrated [20]. The mechanism of microorganism inactivation, beside the thermal effect, could be damaging DNA, RNA, and proteins in microbial cells as occur in UV radiation [18]. The efficacy of these microbial inactivation depends of several operational factors as IR power intensity, food temperature, peak wavelength, and bandwidth of IR heating source; and also depends of the types and physiological phase of microorganisms, and size and type of food materials [21].

12.5.5   OHMIC HEATING (OH)

In the ohmic heating (OH) is well known, the food acts as an electrical resistor and is located between two electrodes, while alternating electric current pass through the circuit. In this case, the electrical resistance causes heat to be generated throughout the food matrix in a uniform and volumetric fashion. OH, is based on the Joule effect, which principle is the dissipation of electrical energy in the form of heat using an electrical conductor, so that electric conductivity of foods is the critical parameter [22] and the needed frequency varies from a few Hz to dozens of kHz, even though, the general electrical system uses frequencies between 50 to 60 Hz. The concept of OH technology is not new and dates to 1897, also known as Joule heating and resistive heating, electro-heating, and electro-conductive heating. Compared to the technologies described above, OH has advantages in heating uniformity, higher temperature in particles than in liquid, reduced fouling, lower cost, and energy efficiency [22–24]. The principal mechanism of microbial inactivation in OH is thermal damage, however, it also may present non-thermal cellular damage attributed to the electric field, which causes the formation of pores in the membrane, increasing permeability, this phenomenon is known as electroporation. A number of studies have been performed to measure the effect of electricity on microorganisms and enzymes in cell suspensions or in food directly under a variety of conditions and most of these are attributed to the inactivation of microorganisms to the thermal effect. With this method, shorter heating times are achieved compared to conventional ones, and in turn avoid the hot surfaces, reducing gradients of temperature. The process parameters such as frequency, the geometry of the cabinet and the material of the electrode are important factors to considered [25]. Among the main applications of OH are blanching, evaporation, thawing, sterilization, pasteurization, pretreatment on mass-transfer operations as extraction and dehydration. This technology has great potential to be used in the process of high-quality products and can be commercially adapted for the aseptic processing of particulate fluid foods.

12.5.6 NONTHERMAL METHODS

In last years, the nonthermal technologies (NTT) have arisen among the emerging procedures to produce safety, fresh-like, convenient, healthy, and high levels of bioactive compounds food commodities. Some of those procedures such as HPP, US, PEF, PL-UV, IOR, and CP have shown high capacity to reduce the microbial counts, inactivate enzymes and maintaining the quality on different food commodities compared to conventional and emerging thermal methods. Moreover, the NTT are presented as highly energy-efficient methods and in consequence, they have gained industrial interest as potential options for the replacement of non-efficient conventional and emerging thermal processing systems used in industry. In addition, NTT can be applied as unique preservative method or also as a combination of different technologies, also called hurdle technology HT. Most of the previously mentioned technologies star to be used and commercialized around the world, being more noticeable their implementation into industries of the first world countries [26]. The efficacy and mechanisms of action of NTT are discussed along this section.

12.5.7 HIGH PRESSURE PROCESSING (HPP)

High pressure processing (HPP) is a recognized and proved method used to extend the shelf-life and to improve the sensory attributes and nutritional aspects, of different solid and liquid food from different origin. In fact, results of microbial inactivation observed in beverages is similar to those achieved by conventional pasteurization. Therefore, high pressure has emerged as a successful method to reduce the microbial counts and thus ensuring the food safety in high moisture liquid or solid foods but never in porous or dried foods. HPP is based in the application of high pressures (100–800 MPa) at low temperature ($< 45°C$). During HPP processing, the pressure is uniformly diffused through isostatic principle from a medium pressure transfer, usually water, to the food products [27]. The effect of HPP is not altered by extrinsic variables such as packaging form and volume of the food [28]. However, the HPP processing variables such as pressure, temperature, and time of treatment as well as the chemical composition and intrinsic parameters of the food can impact in the antimicrobial effect of HPP [29].

As a matter of fact, HPP not only is able to inactivate microorganisms but also the effectiveness to liberate and enhance the amounts of bioactive compounds in fresh fruits has been widely reported in last years [30]. Conversely, some authors have stated that decompartmentalization of cells triggered by HPP results in the liberation of intracellular enzymes such polyphenol oxidase (PPO), which is capable to interact and destroy the phenolic compounds and thus impacting negatively in the antioxidant potential [31, 32]. However, Patras et al. [33] revealed that the content of phenolic compounds, and in turn the antioxidant capacity, is directly related to the residual activity of the PPO after HPP and low pressures and in turn retaining the antioxidant capacity under longer periods of storage.

On the other hand, recent studies [34, 35] point out that HPP provoke decreases of the dielectric constant as well as the deprotonation of charged groups and disruption of bridges and bounds in the cell walls, lead to reduction of bipolar nature of media and increments of permeability, respectively, and thus allowing higher levels in the concentration of phenolic compounds and antioxidant capacity.

12.5.8   ULTRASOUND (US)

In last decades, ultrasound (US) has emerged as a promising alternative for processing in food industry, food research and food analyze. The commercial best-known use of this technology is as a mild treatment to preserve drinks and beverages. In the cavitation phenomena, the sonication prompts the growing and collapse of bubbles that cause turbulence, variation of pressure, shear strength and consequently the cell rupture [36]. Jermann et al. [37] stated that US treatments of high intensity (frequencies below 100 kHz wit energies above 10 W/cm^2) are applied for emulsification, homogenization, extraction, crystallization, dewatering, degassing, and defoaming of biological matrices.

Barba et al. [27] attributed the effects of US to processing variables such as frequency, power intensity and treatment time, but also the intrinsic characteristics of the food such as moisture, density, thickness, as well extrinsic such as the type of radiation, presence of oxygen and absorption capacity has been linked to the its effectiveness [38]. Cavitation mechanism performed by sonication has shown good results in the inactivation of pathogens/spoilage microorganisms in model systems and real foods, however, the antimicrobial effectiveness of US is greatly increased when combined with temperature (thermosonication), pressure (manosonication) or both (manothermosonication) methods [39]. On the other hand, Hernández-Hernández et al. [26] reported that US processing can also alter the quality of the food commodities, increasing the degradation of bioactive compounds as well as affecting the texture, flavor, color, and nutritional value of those products.

12.5.9   PULSED ELECTRIC FIELDS (PEF)

Pulsed electric fields (PEF) is another alternative processing which have shown favorable results over the microbial and enzymatic inactivation and thus extending the shelf life and preserving the organoleptic attributes in different food matrixes, mostly liquid products [40]. PEF treatments implicates the application of high-voltage pulses >0.1 kV/cm for short periods (*ms* or *μs*) to a product located or flowing between two electrodes. In liquids matrixes, PEF can be applied in batch fashion where a small portion of the sample is treated as a unit or well in continuous mode using a pump, which are highly implemented in food processing lines [41]. The effectiveness of PEF treatments is related to processing parameters such as temperature, pulse frequency, width, electric filed strength, polarity, and treatment as well as physicochemical parameters of the media and the microorganism characteristics [42].

Application of PEF treatments leads to the electroporation of the cell membranes, creating aqueous pathways (electropores) that increase the permeability of cell membranes to ions and other molecules [43]. In fact, the electroporation is the principal mechanisms involved in the microbial death. Moreover, interesting data regarding the enzyme inactivation and other food processing-related effects associated to the PEF treatments have been previously reported [44]. According to Elez-Martínez et al. [45], PEF are capable to provoke alterations of the three-dimensional molecular structure of the enzymes, avoiding the fit with active site and preventing the conversion into products and thus, resulting in a reduction of the enzymatic activity. Regarding the effect of PEF on the bioactive compounds, several studies [46, 47] have evidenced the high retention of those compounds after PEF treatments, leading to foods with high nutritional values. However, the antioxidant capacity can be affected in some cases. Again, the treatment intensity, the food physicochemical attributes and the bioactive compound structure, are determining the retention level of the bioactive compounds and the antioxidant capacity.

12.5.10   PULSED LIGHT AND UV

Among light treatments, pulsed light (PL) and ultraviolet (UV) are the two main treatments evaluated in food commodities. The PL processing involve the application a wide range of wavelengths (100–1,200 nm) composed by UV, visible (Vis) and IR lights with the aim of ensure the sanitation of superficies and food commodities. In this case, samples are exposed to pulses of light of high fluency (0.01–50 J cm²) for short periods (from microseconds to tenths of seconds). In the UV treatments cases, samples are subjected to wavelengths exists between X-ray (180 nm) and the VL (400 nm) regions and longer periods (seconds to minutes) than those in PL [48]. Light treatments are generally produced by xenon gas lamps and recently the use of light-emitting diodes (LED) is under study.

The efficacy of PL and UV treatments to inhibit the microbial growth is related to three mechanisms: (1) photochemical effect, largely attributed to the clonogenic death caused by UV radiation but also to damage in the cell wall membranes due to Vis-IR spectrum [49]; (2) photothermal effect refers to the absorption of radiation and increment of the internal temperature up to microbicidal levels [50]; and lastly (3) the photophysical effect have been related to structural changes, deformation of the cell membranes and disruption of the nucleic acids [51]. However, the effect of light treatments is related to (1) processing factors such as fluence, spectral distribution, distance between sample and lamps and number of pulses; (2) physicochemical characteristics of sample such as topography, thickness, transparency, color, and chemical composition [52, 53] and (3) factors linked to the microorganism like type (membrane and cell wall composition and resistance to temperature, pH, and light), density, and production of microbial secondary metabolites [54]. The effects of light treatments over the bioactive compounds on food commodities can be diverse. Light treatments are capable to cause the stress in plant tissues and stimulate the production, synthesis, and liberation of secondary metabolites with potential antimicrobial and antioxidant activity [55]. In general, PL, and UV treatments have shown a great efficacy as a decontamination method, but only few works have been developed to investigate the potential of light treatments as a postharvest treatment to improve the nutritional quality and content of antioxidant and bioactive substances [56].

12.5.11   IONIZING RADIATION (IOR)

The first emerging method applied to replace the canning method and achieve higher safety and healthy characteristics foods was the IOR [57]. Safety of radiation treatments have been largely documented; however, consumers still considering to radiation as controversial and a non-safe method [58]. This preserving method has exhibit great efficacy to inactivate microorganisms and spores in foods using the three principal radiation sources that are electron beam (e-beam), X-ray, and gamma-ray. For *The Codex Alimentarius Commission*, the application of radiation treatments in food commodities is restricted to the use of radionuclides such as 60Co and 137Cs for gamma radiation, and application of maximum energy of 10 MeV and % MeV for generation of e-beam and X-ray, respectively [59–61]. IOR treatments can be applied by direct or indirect methods; in the first one, ionizing particles cause damage in the bacterial DNA and thus avoiding the cells division and replication, while in the second case, the production of oxidizing agents and active molecules such as hydroxyl and hydrogen radicals and hydrated electrons triggered by the interaction between radiation and water molecules, leads to cellular lysis [62]. Nevertheless, effectiveness of IOR is not only affected by the kind of radiation or the application method, but also by the temperature, pH, moisture content, presence or absence of oxygen, the absorbed dose, chemical composition and density and thickness of food [63].

Nowadays, the IOR treatments for commercial food are restricted to the use of doses lower than 10 kGy and 44 kGy to food packages. As a matter of fact, IOR is applied to replace the use of hydrogen peroxide in the sterilizing of packaging materials; however, the authorized doses for food processing are not enough to microbial inactivation so the capacity of IOR to sensitizing spores before others preserving method, still under evaluation. Recent information regarding the effects of radiation over the bioactive compounds and antioxidant capacity of food commodities have demonstrated that IOR is capable to cause conformational and chemical changes in the molecules and thus leading to the formation of molecules with higher stronger antioxidant capacity [64, 65].

12.5.12   COLD PLASMA (CP)

Currently, cold plasma (CP) is an emerging technology applied for sterilization, inactivation of enzymes and functionalization of food but also to altering of physical properties for etching or deposit of thin films in different materials. In fact, because the great potential as sterilization method, CP is the most promising technology for food preservation. In the CP treatments, electromagnetic fields are applied to an inert gas such as He, O_2, N_2 or atmospheric air, to generate a quasi-neutral ionized gas cloud containing reactive molecular species such as electrons, ions, UV photons, atomic species and charged particles. Therefore, CP is capable to degrade proteins, lipids, cellular DNA, but also allows the accumulation of intracellular charged particles which can cause apoptosis, electrostatic disruption, and electroporation of the cells [66–68]. The mechanisms involved in the microbial inactivation by CP are three: (1) chemical interaction between cells and reactive species and charged particles; (2) damage of the membranes and components of the cell wall caused by the UV-radiation (etching), and (3) injure and breakdown of the DNA chains provoked by UV-radiation. Consequently, CP has shown great results in the inactivation of pathogenic/spoilage bacteria, spores, and viruses in food commodities [69–71]. However, the effectiveness of CP to reduce the microbial depends of different processing variables such as feed gas composition, voltage, time of treatment and humidity. On the other hand, Misra et al. [72] attributed the enzymes inactivation to the stability and complexity of enzymes, their sensibility to frees radicals but also by CP processing variables such as inflow of discharging power and mass transfer among liquefied plasma phases. Despite the promising results on the enzymatic inactivation and food sanitation, the effects of CP treatments over the bioactive compounds and antioxidant capacity still under study.

12.5.13 PACKAGING

The preservation methods are indispensable to guarantee the safety as well as the quality attributes; however, packaging techniques are also indispensable to enhance and maximize the preservative function during the storage of commodities. The effectiveness of preservative techniques during storage is greatly depends of the optimization of some factors such as oxygen content, moisture, and WA, light presence, temperature, and microbial contaminants [73]. Currently, the term "packaging" of food products refers to passive (conventional) active (novel) and innovative (intelligent) methods.

Passive packaging (PP) includes the all the traditional ways of packaging such as canning, bagging, boxed, and glass packed. The PP is the most conventional preserving method in which, the packages act as physical barrier to protect the food products from external factors [74]. That kind of packages not only are well known by producers and consumers, but also provide convenience, protection, and product information. However, consumers, and industries demand have prompted the development of functional, innovative, and efficient methods to ensure the food quality for longer periods.

Active packaging (AP) refers to the incorporation of certain additives or "active compounds" into packaging material or within the container systems and are capable to release or absorb substances from the food or surrounding environment to preserve the safe and quality and enhance the shelf life of food commodities. Therefore, AP includes the use of antimicrobials, antioxidants, and biochemical substances that inhibit the deterioration preserving the organoleptic, nutritional, and antioxidant characteristics of food products [75, 76]. On the other hand, the chemical composition of packaging material and the affinity to interact with active compounds also plays an important role in the effectiveness of the AP. In last years, the evaluation of films based on biopolymers seems to be an alternative to replace the conventional non-biodegradable and fossil-derived materials. Active compounds can be added in the packaging material and also as additives (chemical treatments) in the food matrix and inside the packaging space, in liquid and gaseous fashion, respectively [77]. The use of absorbent or releaser pads to control the moisture, flavor or odor inside the packages and the application of nano-systems to encapsulate active compounds to be integrated in packaging materials or as additives in foods has gained importance [78]. Currently, the modification and control of the carbon dioxide, oxygen or to supplement the interior atmosphere with chemical compounds of inside the packaging through AP is the most used method in food industry [79].

Intelligent packaging is a system capable of carrying out functions such as detecting, sensing, recording, tracing, communicating, and applying scientific logic to facilitate the decision to extend shelf life, safety, improve quality, and provide information about products [80]. IP also contributes to identify processes that affect the final food quality. The intelligent packaging uses different tools such as biosensors, gas sensors, printed electronics, chemical sensors, electronic nose, RF, etc., to detect, record, and transmit information pertaining to biological reactions, microbial growth, pH, internal temperature, formation of volatile compounds and headspace composition. Signals generated by sensors are converted into quantifiable electronic response which is used to identify and improve processes and thus optimize the preservation method applied.

12.6   FINAL REMARKS

Food preservations involve the use of different technologies capable to maintain acceptable quality food commodities. The development of new methods responds to the increasing demand of consumers for healthy, nutritive, innocuous fresh-like and functional food products. Effectiveness of thermal and chemical conventional methods to reduce the microbial growth have been strongly documented. However, thermal processing causes the degradation of bioactive and nutritive compounds, while recent studies reveal that chemical compounds (additives) can cause adverse effects on the nutritional and sensory quality of the food products and/or produce toxic compounds. This fact has prompted the study of different emerging technologies to food processing and preservation. The innovative and emerging methods here described not only are useful to achieve safe and high-quality foods by enhancement of bioactive and nutritive compounds, but also are environmentally friendly and effective to optimization of food processes. Additionally, packaging processing plays and important role the stability of the food quality. Active and intelligent packaging has replaced the use of passive methods in order enhance the quality, functionality, and shelf-life of food. Finally, evaluation of nano-systems, sensors, and development of new biodegradable packaging materials is necessary to avoid the food degradation.

KEYWORDS

- **high pressure processing**
- **infrared**
- **microwaves**
- **ohmic heating**
- **radiofrequency**
- **ultrasounds**

REFERENCES

1. Ahvenainen, R., (1996). New approaches in improving the shelf life of minimally processed fruit and vegetables. *Trends Food Sci. Technol., 7*, 179–187.
2. Ahmed, L., Martin-Diana, A. B., Rico, D., & Barry-Ryan, C., (2012). Extending the shelf life of fresh-cut tomato using by-product from cheese industry. *J. Food Process. Preserv., 36*(2), 141–151.
3. Kim, H. J., Fonseca, J. M., Kubota, C., & Choi, J. H., (2007). Effect of hydrogen peroxide on quality of fresh-cut tomato. *J. Food Sci., 72*, S463–7.
4. Tamer, C. E., & Çopur, O. U., (2010). Chitosan: An edible coating for fresh-cut fruits and vegetables. *Acta Hortic., 877*, 619–624.
5. Thumula, P., (2006). *Studies on Storage Behavior of Tomatoes Coated with Chitosan-Lysozyme Films*. PhD Dissertation, McGill University.

6. Jeong, J., Breecht, J. K., Huber, D. J., & Sargent, S. A., (2004). 1-methylcyclopropene (1-MCP) for maintaining texture quality of fresh-cut tomato. *Hort. Science, 39*(6), 1359–1362.

7. Ayala-Zavala, J. F., Oms-Oliu, G., Odriozola-Serrano, I., González-Aguilar, G. A., Álvarez-Parrilla, E., & Martín-Belloso, O., (2008). Bio-preservation of fresh-cut tomatoes using natural antimicrobials. *European Food Res. Technol., 226*, 1047–1055.

8. Augusto, P. E. D., Soares, B. M. C., & Castanha, N., (2018). Conventional technologies of food preservation. Innovative technologies for food preservation. In: Barba, F. J., Sant'ana, A. S., & Koubaa, M., (eds.), *Innovative Technologies for Food Preservation* (pp. 3–23). Elsevier: United Kingdom.

9. Flores-Ramírez, A. J., García-Coronado, P., Grajales-Lagunes, A., García, R. G., Abud, A. M., & Ruiz, C. M. A., (2019). Freeze-concentrated phase and state transition temperatures of mixtures of low and high molecular weight cryoprotectants. *Adv. Polym. Technol.,* 5341242.

10. Fellows, P. J., (2017). Dielectric, ohmic and infrared heating. In: Fellows, P. J., (ed.), *Food Processing Technology* (pp. 813–844). Woodhead Publishing: United Kingdom.

11. Vadivambal, R., & Jayas, D. S., (2010). Non-uniform temperature distribution during microwave heating of food materials: A review. *Food Bioprocess Tech., 3*(2), 161–171.

12. Jiao, S., Luan, D., & Tang, J., (2015). Principles of radio-frequency and microwave heating. In: Awuah, J., Ramaswamy, G. B., & Tang, H. S., (eds.), *Radio-Frequency Heating in Food Processing* (pp. 3–20). CRC Press: Boca Raton, FL.

13. Martins, C. P. C., Cavalcanti, R. N., Couto, S. M., Moraes, J., Esmerino, E. A., Silva, M. C., & Cruz, A. G., (2019). Microwave processing : Current background and effects on the physicochemical and microbiological aspects of dairy products. *Compr. Rev. Food Sci. Food Saf., 18*, 67–83.

14. Tang, J., (2015). Unlocking potentials of microwaves for food safety and quality. *J. Food Sci., 80*(8), E1776–93.

15. Zhao, Y., Flugstad, B., Kolbe, E., Park, J. W., & Wells, J. H., (2000). Using capacitive (radio frequency) dielectric heating in food processing and preservation: A review. *J Food Process Eng., 23*(1), 25–55.

16. Orsat, V., & Raghavan, G. S. V., (2014). Radio-frequency processing. In: Sun, D. W., (ed.), *Emerging Technologies for Food Processing* (pp. 385–398). Elsevier: United Kingdom.

17. Hassan, H. F., Tola, Y., & Ramaswamy, H. S., (2015). Radio-frequency and microwave applications. In: Awuah, J., Ramaswamy, G. B., & Tang, H. S., (eds.), *Radio-Frequency Heating in Food Processing* (pp. 21–30). CRC Press: Boca Raton, FL.

18. Pan, Z., Atungulu, G. G., & Li, X., (2014). Infrared heating. In: Sun, D. W., (ed.), *Emerging Technologies for Food Processing* (pp. 461–474). Elsevier: United Kingdom.

19. Krishnamurthy, K., Khurana, H. K., Soojin, J., Irudayaraj, J., & Demirci, A., (2008). Infrared heating in food processing: An overview. *Compr. Rev. Food Sci. Food Saf., 7*(1), 2–13.

20. Ramaswamy, R., Krishnamurthy, K., & Jun, S., (2012). Microbial decontamination of food by infrared (IR) heating. In: Demirci, A., & Ngadi, M. O., (eds.), *Microbial*

Decontamination in the Food Industry (pp. 450–471). Woodhead Publishing: United Kingdom.

21. Abdul-Kadir, R., Bargman, T. J., & Rupnow, J. H., (1990). Effect of infrared heat processing on rehydration rate and cooking of *Phaseolus vulgaris* (Var. Pinto). *J. Food Sci., 55*, 1472–1473.

22. Goullieux, A., & Pain, J. P., (2014). Ohmic heating. In: Sun, D. W., (ed.), *Emerging Technologies for Food Processing* (pp. 399–426). Elsevier: United Kingdom.

23. Kim, S. S., & Kang, D. H., (2017). Synergistic effect of carvacrol and ohmic heating for inactivation of *E. coli* O157:H7, *S. Typhimurium*, *L. monocytogenes*, and MS-2 bacteriophage in salsa. *Food Control, 73*, 300–305.

24. Samaranayake, C., & Sastry, S., (2014). Electrochemical reactions during ohmic heating and moderate electric field processing. In: Ramaswamy, H. S., Marcotte, M., Sastry, S., & Abdelrahim, K., (eds.), *Ohmic Heating in Food Processing* (pp. 119–128). CRC Press.

25. Jaeger, H., Roth, A., Toepfl, S., Holzhauser, T., Engel, K. H., Knorr, D., & Steinberg, P., (2016). Opinion on the use of ohmic heating for the treatment of foods. *Trends Food Sci. Technol., 55*, 84–97.

26. Hernández-Hernández, H. M., Moreno-Vilet, L., & Villanueva-Rodríguez, S. J., (2019). Current status of emerging food processing technologies in Latin America: Novel non-thermal processing. *Innov. Food Sci. Emerg. Technol., 58*, 102233.

27. Barba, F. J., Ahrne, L., Xanthakis, E., Landerslev, M. G., & Orlien, V., (2018). In: Barba, F. J., Sant'Ana, A. S., & Koubaa, M., (eds.), *Innovative Technologies for Food Preservation* (pp. 25–51). Elsevier: United Kingdom.

28. Martínez-Monteagudo, S. I., & Balasubramaniam, V. M., (2016). Fundamentals and applications of high-pressure processing technology. In: Balasubramaniam, V. M., Barbosa-Cánovas, G. V., & Lelieveld, H. L. M., (eds.), *High Pressure Processing of Food: Principles, Technology and Applications* (pp. 3–17). Springer: New York.

29. Valdramidis, V. P., Taoukis, P. S., Stoforos, N. G., & Van, I. J. F. M., (2012). Modeling the kinetics of microbial and quality attributes of fluid food during novel thermal and non-thermal processes. In: Cullen, P. J., Tiwari, B. K., & Valdramidis, V. P., (eds.), *Novel Thermal And Non-Thermal Technologies For Fluid Foods* (pp. 433–471). Elsevier: United Kingdom.

30. Marszałek, K., Mitek, M., & Skąpska, S., (2015). The effect of thermal pasteurization and high pressure processing at cold and mild temperatures on the chemical composition, microbial and enzyme activity in strawberry purée. *Innov. Food Sci. Emerg Technol., 27*, 48–56.

31. Rastogi, N. K., Raghavarao, K. S. M. S., Balasubramaniam, V. M., Niranjan, K., & Knorr, D., (2007). Opportunities and challenges in high pressure processing of foods. *Crit. Rev. Food Sci. Nutr., 47*, 69–112.

32. Rossle, C., Wijngaard, H. H., Gormley, R. T., Butler, F., & Brunton, N., (2010). Effect of storage on the content of polyphenols of minimally processed skin-on apple wedges from ten cultivars and two growing seasons. *J. Agric. Food Chem., 58*, 1609–1614.

33. Patras, A., Brunton, N. P., Da Pieve, S., & Butler, F., (2009). Impact of high pressure processing on total antioxidant activity, phenolic, ascorbic acid, anthocyanin content and color of strawberry and blackberry purées. *Innov. Food Sci. Emerg. Technol., 10*(3), 308–313.

34. Hurtado-Fernández, E., Gómez-Romero, M., Carrasco-Pancorbo, A., & Fernández-Gutiérrez, A., (2010). Application and potential of capillary electro separation methods to determine antioxidant phenolic compounds from plant food material. *J. Pharm. Biomed. Anal., 53*, 1130–1160.

35. Keenan, D. F., Rößle, C., Gormley, R., Butler, F., & Brunton, N. P., (2012). Effect of high hydrostatic pressure and thermal processing on the nutritional quality and enzyme activity of fruit smoothies. *LWT - Food Sci. Technol., 45*, 50–57.

36. Liu, D., Ding, L., Sun, J., Boussetta, N., & Vorobiev, E., (2016). Yeast cell disruption strategies for recovery of intracellular bio-active compounds — A review. *Innov. Food Sci. Emerg. Technol., 36*, 181–192.

37. Jermann, C., Koutchma, T., Margas, E., Leadley, C., & Ros-Polski, V., (2015). Mapping trends in novel and emerging food processing technologies around the world. *Innov. Food Sci. Emerg. Technol., 31*, 14–27.

38. Bevilacqua, A., Petruzzi, L., Perricone, M., Speranza, B., Campaniello, D., Sinigaglia, M., & Corbo, M. R., (2018). Nonthermal technologies for fruit and vegetable juices and beverages: Overview and advances. *Compr. Rev. Food Sci. Food Saf., 17*, 2–62.

39. Morales-De La, P. M., Welti-Chanes, J., & Martín-Belloso, O., (2019). Novel technologies to improve food safety and quality. *Curr. Opin. Food Sci., 30*, 1–7.

40. Sánchez-Vega, R., Elez-Martínez, P., & Martín-Belloso, O., (2015). Influence of high-intensity pulsed electric field processing parameters on antioxidant compounds of broccoli juice. *Innov. Food Sci. Emerg. Technol., 29*, 70–77.

41. Morales-De La, P. M., Elez-Martínez, P., & Martín-Belloso, O., (2011). Food preservation by pulsed electric fields: An engineering perspective. *Food Eng. Rev., 3*, 94–107.

42. Fox, M. B., (2007). Microbial inactivation kinetics of pulsed electric field treatment. In: Lelieveld, H. L. M., Notermans, S., & De Haan, S. W. H., (eds.), *Food Preservation by Pulsed Electric Fields: From Research to Application* (pp. 127–137). Woodhead Publishing Limited.

43. Chang, D. C., (2012). Structure and dynamics of electric field-induced membrane pores as revealed by rapid-freezing electron microscopy. In: Chang, D. C., Chassy, B. M., Saunders, J. A., & Sowers, A. E., (eds.), *Guide to Electroporation and Electrofusion* (pp. 9–27). Academic Press.

44. Giner, J., Grouberman, P., Gimeno, V., & Martín, O., (2005). Reduction of pectin esterase activity in a commercial enzyme preparation by pulsed electric fields: Comparison of inactivation kinetic models. *J. Sci. Food Agric., 85*, 1613–1621.

45. Elez-Martínez, P., Martín-Belloso, O. M., Rodrigo, D., & Sampedro, F., (2007). Impact of pulsed electric fields on food enzymes and shelf-life. In: Lelieveld, H. L. M., Notermans, S., & De Haan, S. W. H., (eds.), *Food Preservation by Pulsed Electric Fields: From Research to Application* (pp. 212–246). Woodhead Publishing Limited: England.

46. Sánchez-Vega, R., Garde-Cerdán, T., Rodríguez-Roque, M. J., Elez-Martínez, P., & Martín-Belloso, O., (2020). High-intensity pulsed electric fields or thermal treatment of broccoli juice: The effects of processing on minerals and free amino acids. *Eur. Food Res. Technol., 246*, 539–548.

47. Vallverdu-Queralt, A., Odriozola-Serrano, I., Oms-Oliu, G., Lamuela-Raventos, R. M., Elez-Martínez, P., & Martin-Belloso, O., (2012). Changes in the polyphenol

profile of tomato juices processed by pulsed electric fields. *J. Agric Food Chem., 60,* 9667–9672.

48. Zhang, W., & Jiang, W., (2019). UV treatment improved the quality of postharvest fruits and vegetables by inducing resistance. *Trends Food Sci. Technol., 92,* 71–80.

49. Ramos-Villarroel, A. Y., Martín-Belloso, O., & Soliva-Fortuny, R., (2013). Intense light pulses: Microbial inactivation in fruits and vegetables. *CYTA-J. Food, 11*(3), 234–242.

50. Farrell, H. P., Garvey, M., Cormican, M., Laffey, J. G., & Rowan, N. J., (2010). Investigation of critical inter-related factors affecting the efficacy of pulsed light for inactivating clinically relevant bacterial pathogens. *J. Appl. Microbiol., 108,* 1494–1508.

51. Elmnasser, N., Guillou, S., Leroi, F., Orange, N., Bakhrouf, A., & Federighi, M., (2007). Pulsed-light system as a novel food decontamination technology: A review. *Can. J. Microbiol., 53,* 813–821.

52. Koutchma, T., (2019). *Ultraviolet Light in Food Technology Principles and Applications* (2nd edn.). CRC Press: Boca Raton.

53. Uesugi, A. R., Woodling, S. E., & Moraru, C. I., (2007). Inactivation kinetics and factors of variability in the pulsed light treatment of *Listeria innocua* cells. *J. Food Protec., 70*(11), 2518–2525.

54. Fine, F., & Gervais, P., (2004). Efficiency of pulsed UV light for microbial decontamination of food powders. *J. Food Protec., 67,* 787–792.

55. Aguiló-Aguayo, I., Gangopadhyay, N., Lyng, J. G., Brunton, N., & Rai, D. K., (2017). Impact of pulsed light on color, carotenoid, polyacetylene and sugar content of carrot slices. *Innov. Food Sci. Emerg. Technol., 42,* 49–55.

56. Denoya, G. I., Pataro, G., & Ferrari, G., (2020). Effects of postharvest pulsed light treatments on the quality and antioxidant properties of persimmons during storage. *Postharvest Biolol. Technol., 160,* 111055.

57. Roberts, P. B., (2014). Food irradiation is safe: Half a century of studies. *Radiat. Phys. Chem., 105,* 78–82.

58. Farkas, J., Ehlermann, D. A. E., & Mohácsi-Farkas, C., (2014). Food technologies: Food irradiation. In: Motarjemi, Y., (ed.), *Encyclopedia of Food Safety* (pp. 178–186). Elsevier: USA.

59. Ehlermann, D. A. E., (2016). Wholesomeness of irradiated food. *Radiat. Phys. Chem., 129,* 24–29.

60. Feliciano, C. P., (2018). High-dose irradiated food: Current progress, applications, and prospects. *Radiat. Phys. Chem., 144,* 34–36.

61. Ravindran, R., & Jaiswal, A. K., (2019). Wholesomeness and safety aspects of irradiated foods. *Food Chem., 285,* 363–368.

62. Tahergorabi, R., Matak, K. E., & Jaczynski, J., (2012). Application of electron beam to inactivate *Salmonella* in food: Recent developments. *Food Res. Inter., 45,* 685–694.

63. Olatunde, O. O., & Benjakul, S., (2018). Nonthermal processes for shelf-life extension of seafoods: A revisit. *Compr. Rev. Food Sci. Food Saf., 17*(4), 892–904.

64. Gamez, M. C., Calvo, M. M., Selgas, M. D., García, M. L., Erler, K., Bohm, V., & Palozza, P., (2014). Effect of E-beam treatment on the chemistry and on the antioxidant activity of lycopene from dry tomato peel and tomato powder. *J. Agric. Food Chem., 62*(7), 1557–1563.

65. Guerreiro, D., Madureira, J., Silva, T., Melo, R., Santos, P. M. P., Ferreira, A., & Cabo, V. S., (2016). Post-harvest treatment of cherry tomatoes by gamma radiation: Microbial and physicochemical parameters evaluation. *Innov. Food Sci. Emerg. Technol., 36*, 1–9.

66. Liao, X., Liu, D., Xiang, Q., Ahn, J., Chen, S., Ye, X., & Ding, T., (2017). Inactivation mechanisms of non-thermal plasma on microbes: A review. *Food Control, 75*, 83–91.

67. Lunov, O., Zablotskii, V., Churpita, O., Jager, A., Polivka, L., Sykova, E., & Kubinova, S., (2016). The interplay between biological and physical scenarios of bacterial death induced by non-thermal plasma. *Biomaterials, 82*, 71–83.

68. Patange, A., Boehm, D., Bueno-Ferrer, C., Cullen, P. J., & Bourke, P., (2017). Controlling brochothrix thermosphacta as a spoilage risk using in-package atmospheric cold plasma. *Food Microbiol., 66*, 48–54.

69. Baier, M., Ehlbeck, J., Knorr, D., Herppich, W. B., & Schlüter, O., (2015). Impact of plasma processed air (PPA) on quality parameters of fresh produce. *Postharvest Biol. Technol., 100*, 120–126.

70. Li, X., & Farid, M., (2016). A review on recent development in non-conventional food sterilization technologies. *J. Food Eng., 182*, 33–45.

71. Niemira, B. A., & Gao, M., (2012). Irradiation of fluid foods. In: Cullen, P. J., Tiwari, B. K., & Valdramidis, V. P., (eds.), *Novel Thermal And Non-Thermal Technologies For Fluid Foods* (pp. 167–183). Elsevier: United Kingdom.

72. Misra, N. N., Pankaj, S. K., Segat, A., & Ishikawa, K., (2016). Cold plasma interactions with enzymes in foods and model systems. *Trends Food Sci. Technol., 55*, 39–47.

73. Zahra, S. A., Butt, Y. N., Nasar, S., Akram, S., Fatima, Q., & Ikram, J., (2016). Food packaging in perspective of microbial activity: A review. *J. Microbiol. Biotechnol. Food Sci., 6*, 752–757.

74. Brody, A. L., (2000). Smart packaging becomes intellipac registered. *Food Technology Magazine, 54*, 104–106.

75. Dobrucka, R., (2015). Antimicrobial packaging with natural compounds: A review. *Logforum, 12*, 193–202.

76. Wilson, C. T., Harte, J., & Almenar, E., (2018). Effects of sachet presence on consumer product perception and active packaging acceptability - A study of fresh-cut cantaloupe. *LWT - Food Sci. Technol., 92*, 531–539.

77. Biji, K. B., Ravishankar, C. N., Mohan, C. O., & Srinivasa, G. T. K., (2015). Smart packaging systems for food applications: A review. *J. Food Sci. Technol., 52*, 6125–6135.

78. Peighambardoust, S. H., Beigmohammadi, F., & Peighambardoust, S. J., (2016). Application of organoclay nanoparticle in low-density polyethylene films for packaging of UF cheese. *Packag. Technol. Sci., 29*, 355–363.

79. Ahmed, I., Lin, H., Zou, L., Brody, A. L., Li, Z., Qazi, I. M., & Lv, L., (2017). A comprehensive review on the application of active packaging technologies to muscle foods. *Food Control, 82*, 163–178.

80. Medina-Jaramillo, C., Ochoa-Yepes, O., Bernal, C., & Famá, L., (2017). Active and smart biodegradable packaging based on starch and natural extracts. *Carbohydr. Polym., 176*, 187–194.

CHAPTER 13

Strategies During Citrus Waste Utilization: Fermentative Route for Single-Cell Protein Production

ANDREA GUADALUPE FLORES-VALDÉS,[1] GLORIA A. MARTÍNEZ-MEDINA,[1] JOSÉ LUIS MARTÍNEZ-HERNÁNDEZ,[1] ANNA ILINÁ,[1] CRISTÓBAL NOÉ AGUILAR,[1] NATHIELY RAMÍREZ GUZMÁN,[2] and MÓNICA L. CHÁVEZ-GONZÁLEZ[1]

[1]Food Research Department, School of Chemistry, Autonomous University of Coahuila, Saltillo – 25280, Coahuila, México, E-mail: monicachavez@uadec.edu.mx (M. L. Chávez-González)

[2]Center for Interdisciplinary Studies and Research, Autonomous University of Coahuila, Saltillo – 25020, Coahuila, México

ABSTRACT

Using as a referent that essential quantities of wastes are generated in the food industry, novel approaches are needed, mainly directed to develop new management alternatives and treatments that allow their biotransformation into valuable products where biotechnology result in an important tool to achieve this goal. Citric wastes represent an essential and underutilized by-product derived from food industry; where orange gain attention in countries as Mexico, due to their vast popularity, remarkable flavor, and high Vitamin C content, the productive chain result in waste with high peels volume and fibers, and they could be exploited using fermentative process. The fermentative process employs a set of living organisms to generate a novel product or derived products. Since ancient times, the microorganism has been exploited by a human, nevertheless their intensification and novel approaches development is demanded. This chapter aims to explore the orange waste in Mexico and its prospective use as a substrate in the fermentative process for single-cell protein production.

## 13.1	INTRODUCTION

The presence of highly variable regions, climates, and lands are some factors that make Mexico a country with vast natural wealth and therefore allow the production of different fruits and vegetable species, both for national and international supply.

The primary agriculture production in Mexico during 2017 is represented by fruits [97] and ranks the position number five, behind the former producer leaders: China, Brazil, United States, and India [1].

Citrus fruits constitute one of the most relevant Mexican fruit crops related to the employed area for its cultivars and parallel to the jobs generated from activities linked to production, industrialization, and marketing at the national or international level. This activity is carried out in regions with tropical and sub-tropical climates with major domestic production is placed in five states: Veracruz, San Luis Potosí, Tamaulipas, and Nuevo Leon [1].

The most common citrus fruits are represented by orange (*Citrus sinensis*), mandarin (*Citrus reticulata*), lemon (*Citrus limon*), grapefruit (*Citrus paradisi*) and lime (*Citrus aurantifolia*), usually characterized by spectacular flavor and apport to human nutrition with components as Vitamin C [2].

Citrus fruits despite to be a flavory and important ingredient source during human feeding, the quantities of fresh fruit consumed are scarce, and high amounts are processed; where only the near to 50% on fruit weight is util during processing techniques [2] and as a consequence, these activities, generate a critical source of by-products with environmental or health implications when are not treated or exploited in correct ways.

The chapter encompass the evaluation of potential and possible route to follow for citrus wastes employment as a raw material in biotechnological fields for single-cell protein production, where first analyze the citrus components, the treatments needed for their employment, the biotechnological tool, and their most evident influence parameters.

13.2 CITRUS FRUITS COMPOSITION

The entire citrus composition possesses molecules with exciting properties; these compounds are distributed in different structures. The main structures are displayed in Figure 13.1. The analysis of compounds that could be found in citrus fruits is vital because we can focus the extraction or enhancement strategies or anticipate the compounds which possible hinder or promote the citrus by-products employment in certain areas.

Different authors explore the main components in citrus fruits, for example, the epicarp represents the external portion in husk also known as flavedo, in this part carotenoids are concentrated, and also we can find oleaginous glands loaded with essential oils from each citrus specie [3, 4], one the other hand, the mesocarp is located directly under epicarp and also named albedo (Figure 13.1), this is a fluffy coat with variable thickness depending on different citrus fruits with a particular cellular conformation with cellulose and hemicellulose substances, the mesocarp covers the edible portion in fruits, and both (flavedo and albedo), conforms pericarp commonly named crust or fruit peel; the endocarp is the name given to the edible part in citrus fruits and is structured by vesicles where the juice is extracted [4, 5].

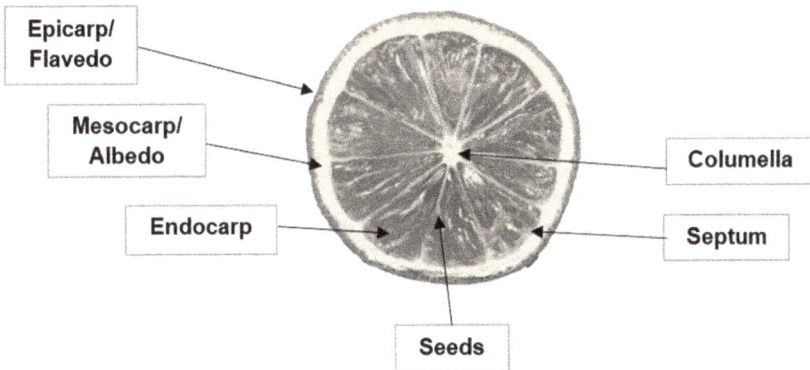

FIGURE 13.1 Classical citrus fruit structure.

Critics are generally directed to juice production, generating a waste principally constituted by seeds, endocarps, and mesocarps, identified as lignocellulosic materials. The fundamental fibers constituents present in citrus are described in subsections.

13.2.1 CELLULOSE

Over the time, cellulose represents one of most relevant carbohydrates, recognized as a most abundant and renewable resource in nature at worldwide level, and one of most abundant compound in cell wall and close correlated to other components as hemicellulose, lignin, and pectin and additionally Cellulose is recognized as a group of linear polysaccharides chains with more than 2000 D-(+) glucose unit, with β-1,4-glucosidic linkage, these chains interact with each other by hydrogen bonds resulting in microfibers with highly ordered regions conferring characteristics as insolubility, rigidity, and resistance to chemical or enzymatic substances, however when the hydrolysis is carried out characteristic dimers as cellodextrin or cellobiose and monosaccharides as glucose was obtained [6].

13.2.2 HEMICELLULOSE

The hemicellulose is a ramified polysaccharide and is considered as a second natural polymer after cellulose utilized for renewable energy [7–9], the hemicellulose content cellulose microfibers; also, is considered a heteropolymer conformed by a pentose (D-xylose and L-arabinose) and hexoses (D-mannose and D-galactose) [10]. This structure contents near 200 monosaccharide units joined by β-1-4 and β-1, glucoside, amorphous, and highly branched, making it a highly soluble compound during thermochemical procedures [11], the classical thermal hemicellulose solubilization is raised at 150 ªC under neutral conditions and at 120°C under acid conditions [12].

13.2.3 PECTIN

Pectin, as hemicellulose, constitute a heterogeneous polymer composed by galacturonic acid methyilic ester, joined by α-1,4 linkage between galacturonic acid and neutral sugars as rhamnose, galactose, and arabinose [13].

13.2.4 LIGNIN

Lignin is a tridimensional polymer composed of phenylpropanoids (p-hidroxyfenil, guiacyl, and syringil) joined by ester and carbon-carbon bonds [14], also has been reported cinnamic acid units (p-coumaric, ferulic, and sinapic). A few in a carbohydrate-lignin bridge system [15], in natural fruit structure, the lignin content helps to protect the cellulose from microbial attack, confers impermeability and maintains the cellulosic fiber adhered [16].

The lignin is a hydrophobic compound, a characteristic frequent in aromatic compounds, for this reason, it cannot be dissolved into aqueous media, unless polar substituents will be added trough chemical reactions with the function to stabilize the lignin into water [16].

The lignin contents represent a challenge during citrus biomass conversion into added-value products, especially when living organism intervenes.

Chiramonti et al. [16] argues that is necessary remove lignin to preserve polysaccharides as hemicellulose and cellulose due to that the formerly mentioned structures, represent the main source of reducing sugars and as a consequence, the degradation and conversion of vegetal tissue to fermentable sugar major indicative; is essential to mention that each citrus fruit is different and due to, their lignin contents fluctuate, e.g., lemons wastes possess low lignin percentages making it suitable as raw material during bio-conversion process as bioethanol production [17].

13.2.5 ARABINOGALACTAN (ARGA)

Arabinogalactan (ArGa) are widely represented in the vegetal kingdom, below we describe different classifications:

> **Type I:** Present a main chain with β-D-galactopyranose with β-(1-4) linkage 18, substituted with α-L-Arabinofuranosyl (L-Araf) short chains, generally in the third position by (1-5) linkage. The α-L-Araf units could be until 20–40% of total monosaccharides in the molecule the ArGa´s type is generally associated with citrus and apple pectin [19].

> **Type II:** Know as 3,6-Arabinogalactan, possess a main chain formed by D-galactopyranose with β-(1-3) linkage 20, substituted

in the sixth position by mono- and oligo-saccharides chains integrated by arabinose and galactose, they are more frequent than the ArGa Type I, generally associated with proteins (ArGaP), joined by covalent linkages to peptides mostly composed by hydroxyproline [21–23].

13.2.6 XYLAN

Xylan possesses importance due to the structural resistance in the cell wall that provided by covalent and no-covalent bonds generation between xylan and other cell wall components [24]. Xylan is a no-cellulosic structural polysaccharide with elevated abundance in angiosperms, representing to 20–30% in dry weight, and composed by D-xylopyranose units with β-(1-4) linkage with shorth branching; the branching could contain glucuronic acid or arabinose units or even oligosaccharide integrated by xylose, arabinose, galactose or glucose [25], the frequency and branching depend on xylan source [26–28], has been reported non-branched xylan composed exclusively by xylose units in some algae, in this case with the presence of β-(1-4) and β-(1-3) linkage, and with structural and as reserve element as possible biological finalities [29].

The aforementioned group of molecules are present in the citrus waste, the analysis of their normal distribution, presence, and functionality supplies a guideline for their exploitation, nevertheless for biotechnological approach in occasions is necessary a set of strategies as pretreatments for making feasible the living organisms utilize the molecules present in citrus or other agro-wastes, the pretreatment strategies are discussed in further sections.

13.3 PRETREATMENTS

The pretreatment strategies are an indispensable step during citrus waste processing, where their natural composition needs some modifications making possible to obtain a fermentable sugars rich material, in fact, in nature cellulose degradation is highly crucial in the carbon cycle and is carried out by soil microorganism as fungi or bacteria, that employs their enzymatic tools to break the polymeric components and generate diverse carbohydrates which could be used as energy source [30, 31], this natural process exemplifies one strategy which can be escalated for citrus residues achievement, nevertheless below evaluate other intends of action that could be employed, modified or combined.

13.3.1 CHEMICAL HYDROLYSIS

The chemical hydrolysis is commonly carried out using acids, employing concentrated sulfuric or hydrochloric acid, with disadvantages as their toxicity and corrosivity; also requires corrosive resistant material-reactors, generating an expensive process, even for making the project economic-viable is necessary to recover the acids from reaction media [32].

13.3.2 PHYSICOCHEMICAL PRETREATMENT

The physicochemical treatment includes technology that allows the ligno-cellulosic material disintegration combining different approaches, here we can find the steam explosion or radiation methods employing ultra-sound (US) or microwaves (MWs), this substitutes the classic chemical substances integration to the system for another kind of energetical forces, nevertheless demand a set of specialized technology [33].

13.3.3 ENZYMATIC HYDROLYSIS

The enzymes are protein catalyzers generated by a living organism, which main purpose is to accelerate highly specific reactions, during the organism's development [34], industrial enzymes are used to propitiate the usage of a new class of raw materials or to alter the physical properties for most efficiently proceeding, generating an increase in solubility, lowering their viscosity, changing color, odor, texture, flavor, or useful life among other features [35, 36].

During enzymatic hydrolysis, the cellulases break the chains from the cellulose and hemicellulose. In this rupture, the cellulose breaks down into glucose, and the hemicellulose breaks down into monosaccharides such as xylose, arabinose, galactose, and mannose, among others. This type of enzymatic treatment has the advantage of not presenting corrosive effects, is a highly specific procedure and carried out under soft reaction conditions.

The most common process consists on mix the enzyme dissolution with the lignocellulosic substrate, controlling parameters like pH, temperature, and homogeneity during the process, but also considering the type of substrate when this kind of factor are determined the optimization could be executed [32].

How above was mentioned the vegetal lignocellulosic polymers degradation in nature is determinant in the carbon cycle, ensuring the return of nutrient to soil assisting the environment sustainability [37], which extensively depends on lignocellulosic enzymes produced by different organisms including superior organisms and microorganisms like fungi and bacteria which normally secrete an enzyme group with the primary purpose to degradation the vegetal membranes [7]. The enzymes that are generally used for lignocellulosic degradation are discussed in subsections.

13.3.3.1   CELLULASES

Cellulase enzymes are produced by a variety of fungi and bacteria with aerobic, anaerobic, mesophilic, and thermophile characters; however just a few of them possess the capacity to excrete these enzymes to the media [38], the cellulolytic system act generally in a pH value between 4–6, they are thermostable and could act even up 60°C [39].

The cellulolytic systems derived from microorganisms possess the particularity to present synergism as plan of action to disrupt lignocellulose wastes where enzymes are implicated, this action are described in further sections.

13.3.3.2   ENDO-1-4-B-GLUCANASES OR 1,4-B-GLUCAN-HYDROLASE (EC 3.2.1.4)

The endoglucanases break in an aleatory form the cellulose chains with β-(1-4) linkage, and generate oligosaccharides with variable extension and new reducing extreme, acting in the amorphous zone in the molecule and performing a decrease in polymerization grade with glucose, cellobiose, and cello triose as products [40].

13.3.3.3 EXO-β-GLUCANASES OR 1,4-B-GLUCAN CELLOBIO-HYDROLASES (EC 3.2.1.91)

The exo-glucanases or cellobiohydrolases act in a cellulose extreme releasing glucose and cellobiose (glucose disaccharides), classified in cellobiohydrolases I (CBHI) working in reducing extreme and cellobiohydrolases II (CBHII), acting in not-reducing extreme [40].

13.3.3.4 β-GLUCOSIDASES OR CELLOBIOSES (EC 3.2.1.21)

The β-glucosidase hydrolyze cellobiose and short-chain products as cellodextrines into glucose monomers [40]. The enzymatic method is environmentally friendly and increases the lignocellulosic material accessibility without generation of high toxicity wastes, anyhow the industrial process remains limited due to the high time demanded [41], in addition to this, is necessary contemplate a group of factors with strong influence over the process efficiency, as substrate physical characteristics, substrate composition, polymerization grade in biomass and multienzyme complex synergy [42].

13.3.3.5 LIGNINOLYTIC ENZYMES

The lignin degradative enzymes are named ligninase. They are oxidative unspecific acting through a non-specific mediator in contrast to the hydrolytic proteins as cellulases or hemicellulose [43]. The main ligninolytic enzymes are the manganese peroxidase (MnP) and the lignin-peroxidase (LiP) that catalyze a variety of oxidative reactions H_2O_2 dependents and the laccases that oxidize phenolic compounds reducing molecular oxygen to water [43]. During lignin degradation, also participates, an extracellular hydrogen peroxide generator enzyme as glyoxal oxidase or glucose oxidase, which product is essential for peroxidase activities [37]. The MnP and Laccases could oxide the phenolic structure in lignin while MnP could degrade the lignin in an effective way due to their significant reduction potential and eventually their CO_2 release [44], in other hand the LiP efficiently oxidize the non-phenolic structure in lignin when is compared with the other enzymes [45], if well the laccase is not the most specific enzyme during lignin breakdown, this enzyme collaborate as important effector along additional mediators, also has been reported that the pH action range for exploit their maximum potential varies between each enzyme where pH near to 4.5 for LiP, from 4.5 to 5.0 for MnP and from 3.0 to 5.7 for laccase explains the lignin degradation phenomena complexity [45], recently other ligninolytic enzymes has been discovered, as versatile peroxidase from white decaying Basidiomycetes, this enzyme combines the characteristics from MnP and LiP related to substrate specificity oxidizing ability against phenols, hydroquinone or dyes with variable redox potential [43, 46] also has been reported some MnP, LiP, and Laccases produced extracellular by fungi [43, 47].

13.3.3.6   HEMICELLULOSE AND XYLAN DEGRADATIVE ENZYMES

As hemicellulose and cellulose possess a very related structure, the enzymes which could be implicated in this kind of polysaccharide hydrolysis are very similar [7, 8] when hemicellulose is degraded until monomeric carbohydrates and acetic acid, and as a consequence xylan is obtained in significant proportions, therefore, the presence of xylanolytic enzymes is necessary [8].

Xylanase is typically extracellular enzymes and regularly act in two different approaches: first, breaking the β-1,4 linkage in D-xylopyranose that form the glycosidic link with hemicellulose polymer and second degrading the branch joint, triggering a special re-arrangement enabling the enzyme-substrate complex interaction [8, 11].

Has been reported fungi, bacteria, plant, and insects as xylanase producers [11, 48], the filamentous fungi represent one of most studied xylanolytic enzymes producers [48], in accordance to CAZy database (http://www.cazy. org/fam/acc_GH.html) exist to leading xylanase group with different action approach the GH10 xylanase family with a flexible specificity to xylan degradation but could degrade some substitutes and the GH11 xylanase family which is highly specific to xylan acting only over this substrate [11].

13.3.4   SUBPRODUCTS IN LIGNOCELLULOSIC BIOMASS HYDROLYSIS

How was explained before, and Figure 13.2 displays, after lignocellulosic residues breakdown the main product formed are: glucose, xylose, galactose, and phenolic compounds, however from monomers produced, some of them could suffer a decomposition during the treatment, as 5-hydroxy methyl furfural (HMF) derived from hexoses and furfural from pentose, but also could generate levulinic acid, formic acid or acetic acid [49], also has been reported that this kind of compounds could be affecting the microorganism performance presenting a slow growth in some strains or low biomass generation, also has been reported that some yeast could metabolize the furfural compounds, in the other hand the acids could affect intracellular pH in microorganism making harder that the organism could exert their biochemical tool efficiently [50].

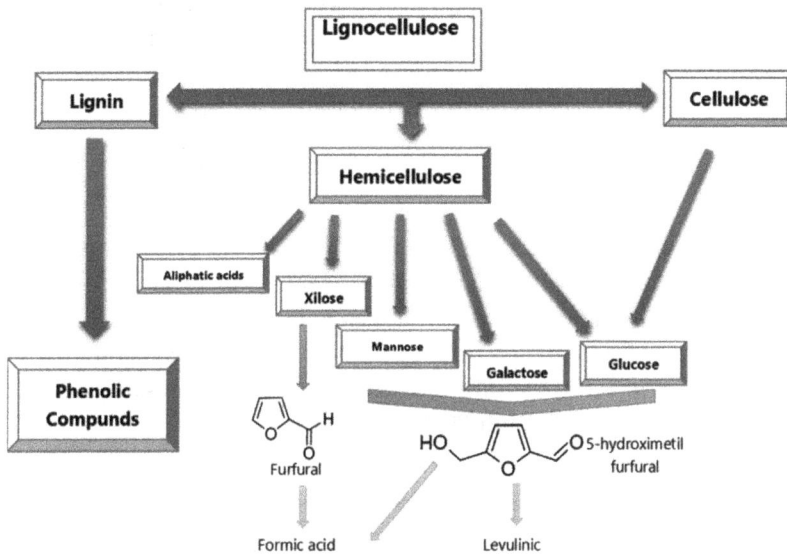

FIGURE 13.2   General single-cell protein production diagram.

13.4   FERMENTATIVE PROCEDURES

Other biotechnological tools that could be executed during lignocellulosic wastes exploitation correspond to fermentative procedures; an instrument that has been utilized since ancient times, that propose the use of different living organisms for materials transformation or valuable products production, nevertheless their finality has been adjusted to human demands in diverse concerning topics as food, chemical, and pharmaceuticals fields.

The fermentative procedure uses a wide range of microorganisms like algae, bacteria, yeast or filamentous fungi [51] and is generally classified in two different types: solid-state fermentation (SSF) or submerged fermentation (SMF) also known as liquid fermentation (SmF) [52], the main difference in both procedures is related to the water quantities involved in each process; the SSF is developed in a substrate with only the moist enough to support the microorganism growth [53–55], while the SMF involves a microorganism that growths in a fully aqueous media [56, 57].

For SSF, two kinds of cultures have been recognized, those where the not-water soluble solid material properly moisten act as nutrients source and support for microorganism development versus those, where the substrate is fortified with nutrients for the improvement of the crucial metabolite generation [58, 59], this kind of fermentation process has been reported as a suitable choice for microbiologic products obtention during mushroom production, novel animal feeding development or enzyme production, among others, where is important to highlight the possibility to incorporate agro-industrial residues in these procedures [52], otherwise while SmF is employed insignificant proportion for bacteria and yeast [60], also has been developed for a significant fungi variety, offering the production of enormous biomass volumes in less time [61].

The class of fermentation utilized in the process depends on the result demanded, but in Table 13.1 are described some advantages and disadvantages in general for both methods.

TABLE 13.1   Comparative Scheme between SSF and SmF

Solid State Fermentation	Submerged Fermentation
Advantages	
• Culture media simplicity	• Media could be designed
• Less spatial demand in bioreactors	• Easier parameter control
• Less water utilization and effluents generation	• Major nutrients availability
• Higher yields	• Simple instrumentation utilized during analysis
• Reduced contamination risks	• Easier metabolic heat dissipation
• Lower sterilization requirements	• Lower variability
• Similar nature-environment development (fungi)	• Easier microbial biomass separation
	• Great scaling probability

TABLE 13.1   *(Continued)*

Solid State Fermentation	Submerged Fermentation
Disadvantages	
• Substrate's pretreatment claim	• Elaborated downstream process
• Parameters with high control difficulty	• High liquid waste generated
• Complex analytic methods for analyze and control the procedure	• High energy consumer
	• Difficult usage in Filamentous fungi
• High inoculum volume requirement	• Costly ingredients
• Poor agitation	• Contamination problems
• Lack bioreactor designs and engineering process in large scale	

Source: [54, 55, 57, 58, 60, 75, 98–100].

13.4.1   CONDITIONAL VARIABLES DURING FERMENTATION

If well, both processes possess equal importance, each procedure is adjusted to a user and expected results needs. However, we can talk about a group of general parameters involved in the fermentation control.

13.4.1.1   MOISTURE

Without a doubt the moisture during SSF is a crucial factor and is directly related to WA, influencing growth and metabolic activity in microorganisms which is extensively documented [62–67], this parameter is a decisive choice due to that determine the class of microorganism available to growth in a substrate which is in function to the type of support and their physic-chemical characteristics.

The microscopical fungi represent the ideal organisms for growth in SSF despite to their WA requirements, also is essential to mention that the criteria for metabolite production are superior that the needed for growth, remarking the importance in this parameter control [63, 68, 69], in the other hand an excess in the Aw value could promote low porosity in a substrate and as a consequence in oxygen diffusion allowing bacterial contamination or aerial mycelia formation [70], while smaller water in the system could advocate phenomena as sporulation; and generally the system present water losses due to the microorganism metabolic activity

[69], for this reason, is recommended the frequent humidifiers usage or sterile water [59].

13.4.1.2 TEMPERATURE

The temperature constitutes an environmental factor that contributes relevantly in microbial growth and products formation [71], also generate influence in cellular metabolism, potential, and stability in cellular components as proteins (enzymes) where we can mention reaction velocity, molecular interactions and other physicochemical properties [72, 73], the incubation temperatures is guided by the optima growth temperature, tough this not guarantee alternative metabolites generation [74] during SSF is expected an increase in system temperature along the process, especially in intern substrate zone affecting directly in growth, germination, and sporulation, making significative in bioreactor design include mechanisms that allow the heat dissipation as control jackets or staying in controlled temperature room [69].

13.4.1.3 AERATION AND AGITATION

Trough aeration and agitation in fermentative process we are trying to enhance the mass transfer process at inter- and intra-particle levels, during interparticular procedure the most crucial mass transfer phenomena is the gas diffusion, specially oxygen (aerobic process) and during SSF depends on porosity proportion in solid material and the aeration, the cavity proportion is given by particle geometry, moister content and substrate chemical nature that need to be represent by about of 30% of total fermentation volume [75] is especially essential increase the oxygen transfer and CO_2 elimination by aeration in interstitial [75], for SmF the agitation provides this benefits and also promote de distribution and interaction between cells and media components [76, 77], for this reason they could affect in metabolism [78], where the poor agitation rates could provoke gradients that could be relevant in growth and metabolite production or even when is exceeded in the case of fungal fermentation could cause fungal structure damage or morphologic changes [51, 79].

13.4.1.4 PH

During microbial process the pH media is considered one of most valuable factor that could reverberate in cellular development and also in products formation, the pH value could influence and change the metabolic fluxes, if well, the environmental demands changes between strains [80], the microorganism could growth in a wide pH range adapting their genic expression [81], avoiding high intracellular pH variations with the finality to maintain fundamental process as DNA transcription, protein synthesis and enzymatic activities [82], nevertheless this process demands high energetic maintenance and as consequence generate influence in metabolite production or biomass generation [80], the medium pH affects strongly enzymatic procedures and transport across membranes thus an optimum valor could promote a high metabolic efficiency [71, 83], al pH could affect fungal morphology for this reason is essential to ensure strategies for control them obtaining a highly efficient productivity [84], the SmF demand a very efficient pH control while SSF is more flexible due to the tampon capacity in substrates [85] however occasionally is needed the use of buffered solution for moisture the material support towards that avoid the pH changes in localized area, this strategy result adequate in cultures where the nitrogen source is added as ammonium salts due to the severe pH changes that provoke [85].

13.5 SINGLE CELL PROTEIN (SCP) PRODUCTION

Protein intake during human nutrition represent a key ingredient, due to that after their consume provide essential elements necessary to energy intake and construct crucial molecules related to a proper body function 86, in accordance to WHO the daily protein intake for adult women and man is 0.83 g/Kg and is adjusted to different ages and health issues [87], nevertheless owing to the fast population growth and their future projections the food insurance is on risk, further the obtention by traditional sources implies in occasions high environmental demands or low quality products, being a priority to achieve a set of strategies that could generate novel food sources, where one strategy that could be employed for grant this macronutrient is the microbial derived proteins production [88], in general, the microbial biomass represents a source of a wide relevant products due to their fundamental composition: carbohydrates, lipids, proteins, nucleic acids, vitamin, among others [89] the microbial protein also named single cell protein (SCP) could be generated from different microorganism as algae, bacteria or fungi (filamentous and yeast) [90], the microbial biomass derived from yeasts is considered as an alternative protein source with capability to satisfy the worldwide food demand, during an actual scene with low agriculture production and additional to tendency to population growth, giving a potential alternative for ensure the food security and human nutrition which constitute the mains concerning in food industries and scientific community [91], around world, unicellular protein is used as an important protein concentrate for animal feed, soil additive and flavoring for human consume [92], the unicellular protein obtention involve numerous biochemical process variety, where diverse organism could be exploited [93] represented roughly in Figure 13.3, these representation remark that during this biotechnological procedure is demanded a carbon source where different lignocellulosic materials could be employed, including citrus waste, materials that normally are nitrogen and mineral deficient expecting a supplementation [94, 95].

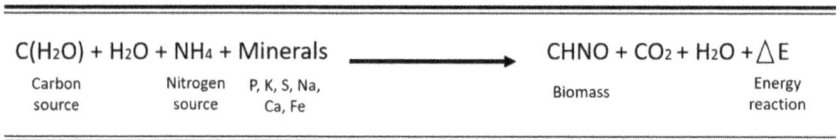

$$C(H_2O) + H_2O + NH_4 + Minerals \longrightarrow CHNO + CO_2 + H_2O + \triangle E$$

| Carbon source | Nitrogen source | P, K, S, Na, Ca, Fe | | Biomass | Energy reaction |

FIGURE 13.3   Biomass formation from different carbon source.

For SCP production the biotechnology exploits the fermentative procedures, generating advantages as generating protein in a short time and small space, nevertheless, the scientific community is struggling a set of parameters that need to be controlled for enhancing the productivity and quality especially in carbon and nitrogen source in addition to the above classic fermentative parameters mentioned [96].

13.6   CONCLUSIONS

The lignocellulosic biomass conversion is realized in different steps, including pretreatment allowing the natural structures rupture, the total or partial solubilization of lignin and hemicellulose and the cellulose crystallinity reduction after that the second stage involve the enzymatic hydrolysis endorsing the saccharification in media for arising with the third stage the fermentative process for obtaining different valuable products, each lignocellulosic material as citrus fruits wastes demand a personalized strategy for exploiting them for the extraordinary structural composition. The implementation of single-cell protein production explores one crucial issue.

ACKNOWLEDGMENT

The authors would like to acknowledge financial support from AGFV to UACOAH-PTC-421 and GAMM to CONACyT for the scholarship with number: 750511, and the Food Research Department in Autonomous University of Coahuila, especially the Bioprocess and Bioproducts Group and Nanobioscencie Group.

KEYWORDS

- **arabinogalactan**
- **biotechnology**
- **citrus waste**
- **lignin-peroxidase**
- **manganese peroxidase**
- **single-cell protein**

REFERENCES

1. Sandoval, K. V., & Ávila, D. D., (2019). Citrus in Mexico: Technical efficiency analysis. *Anal. Económico., 87*, 269–283.
2. Torres-León, C., Ramírez-Guzmán, N., Londoño-Hernandez, L., Martinez-Medina, G. A., Díaz-Herrera, R., Navarro-Macias, V., Álvarez-Pérez, O. B., et al., (2018). Food waste and byproducts: An opportunity to minimize malnutrition and hunger in developing countries. *Front. Sustain. Food Syst., 2*, 52.
3. Ting, S. V., & Attaway, J. A., (1971). Citrus fruits. In: Hulme, A. C., (ed.), *The Biochemistry of Fruits and Their Products* (pp. 107–179).

4. Ting, S. V., & Rouseff, R. L., (1986). *Citrus Fruits and Their Products: Analysis, Technology.* Dekker Marcel, Dekker: New York, USA.

5. Rost, T. L., Barbour, M. G., Thornton, R. M., Weier, T. E., & Stocking, C. R., (1979). Botany: A brief introduction to plant biology. In: Sons, J. W., (ed.), *A Brief Introduction to Plant Biology.* Wiley: New York.

6. Brett, C. T., (2000). Cellulose microfibrils in plants: Biosynthesis, deposition, and integration into the cell wall. In: *International Review of Cytology* (Vol. 199, pp. 161–200) Academic Press: New York.

7. Malherbe, S., & Cloete, T. E., (2002). Lignocellulose biodegradation: Fundamentals and applications. *Rev. Environ. Sci. Biotechnol., 1,* 105–114.

8. Sánchez, C., (2009). Lignocellulosic residues: Biodegradation and bioconversion by fungi. *Biotechnol. Adv., 27,* 185–194.

9. Saratale, G. D., & Oh, S. E., (2012). Lignocellulosics to ethanol: The future of the chemical and energy industry. *African J. Biotechnol., 11,* 1002–1013.

10. Saha, B. C., (2000). α-L-arabinofuranosidases: Biochemistry, molecular biology and application in biotechnology. *Biotechnol. Adv., 18,* 403–423.

11. Moreira, L. R. S., & Filho, E. X. F., (2008). An overview of mannan structure and mannan-degrading enzyme systems. *Appl. Microbiol. Biotechnol., 79,* 165.

12. Hendriks, A., & Zeeman, G., (2009). Pretreatments to enhance the digestibility of lignocellulosic biomass. *Bioresour. Technol., 100,* 10–18.

13. Voragen, A. G. J., Coenen, G. J., Verhoef, R. P., & Schols, H. A., (2009). Pectin, a versatile polysaccharide present in plant cell walls. *Struct. Chem., 20,* 263.

14. Banoub, J. H., Benjelloun-Mlayah, B., Ziarelli, F., Joly, N., & Delmas, M., (2007). Elucidation of the complex molecular structure of wheat straw lignin polymer by atmospheric pressure photoionization quadrupole time-of-flight tandem mass spectrometry. *Rapid Commun. Mass Spectrom. An Int. J. Devoted to Rapid Dissem. Up-to-the-Minute Res. Mass Spectrom., 21,* 2867–2888.

15. Aro, N., Pakula, T., & Penttilä, M., (2005). Transcriptional regulation of plant cell wall degradation by filamentous fungi. *FEMS Microbiol. Rev., 29,* 719–739.

16. Chiaramonti, D., Prussi, M., Ferrero, S., Oriani, L., Ottonello, P., Torre, P., & Cherchi, F., (2012). Review of pretreatment processes for lignocellulosic ethanol production, and development of an innovative method. *Biomass and Bioenergy, 46,* 25–35.

17. Grohmann, K., & Baldwin, E. A., (1992). Hydrolysis of orange peel with pectinase and cellulase enzymes. *Biotechnol. Lett., 14,* 1169–1174.

18. Hori, M., Iwai, K., Kimura, R., Nakagiri, O., & Takagi, M., (2007). Utilization by intestinal bacteria and digestibility of arabinogalactan from coffee bean *in vitro. Japanese J. Food Microbiol., 24,* 163–170.

19. Ebringerová, A., Hromádková, Z., & Heinze, T., (2005). Hemicellulose. In: Heinze, T., (ed.), *Polysaccharides I: Structure, Characterization and Use* (pp. 1–67). Springer: Berlin Heidelberg.

20. Hinz, S. W. A., Verhoef, R., Schols, H. A., Vincken, J. P., & Voragen, A. G. J., (2005). Type I arabinogalactan contains β-d-Galp-(1→3)-β-d-galp structural elements. *Carbohydr. Res., 340,* 2135–2143.

21. Clarke, A. E., Anderson, R. L., & Stone, B. A., (1979). Form and function of arabinogalactans and arabinogalactan-proteins. *Phytochemistry., 18,* 521–540.

22. Fincher, G. B., & Stone, B. A., (1974). A water-soluble arabinogalactan-peptide from wheat endosperm. *Aust. J. Biol. Sci., 27,* 117–132.
23. Redgwell, R., & Fischer, M., (2006). Coffee carbohydrates. *Brazilian J. Plant Physiol., 18,* 165–174.
24. Thomson, J. A., (1993). Molecular biology of xylan degradation. *FEMS Microbiol. Lett., 104,* 65–82.
25. Rodríguez-Palenzuela, P., García, J., & De Blas, C., (1998). Fibra soluble y Su Implicación En Nutrición Animal: Enzimas y Probióticos. Avances en nutrición y alimentación animal, *14,* 227–240.
26. Aspinall, G. O., (1980). Chemistry of cell wall polysaccharides. In: *Carbohydrates: Structure and Function* (pp. 473–500). Elsevier.
27. Scheller, H. V., & Ulvskov, P., (2010). Hemicelluloses. *Annu. Rev. Plant Biol., 61,* 263–289.
28. Gullon, P., González-Muñoz, M. J., Van, G. M. P., Schols, H. A., Hirsch, J., Ebringerova, A., & Parajó, J. C., (2010). Production, refining, structural characterization and fermentability of rice husk xylooligosaccharides. *J. Agric. Food Chem., 58,* 3632–3641.
29. Yamagaki, T., Tsuji, Y., Maeda, M., & Nakanishi, H., (1997*)*. NMR spectroscopic analysis of sulfated β-1, 3-xylan and sulfation stereochemistry. *Biosci. Biotechnol. Biochem., 61,* 1281–1285.
30. Gefen, G., Anbar, M., Morag, E., Lamed, R., & Bayer, E. A., (2012). Enhanced cellulose degradation by targeted integration of a cohesin-fused β-glucosidase into the *Clostridium thermocellum* cellulosome. *Proc. Natl. Acad. Sci., 109,* 10298–10303.
31. Coradetti, S. T., Craig, J. P., Xiong, Y., Shock, T., Tian, C., & Glass, N. L., (2012). Conserved and essential transcription factors for cellulase gene expression in ascomycete fungi. *Proc. Natl. Acad. Sci., 109,* 7397–7402.
32. Binod, P., Janu, K. U., Sindhu, R., & Pandey, A., (2011). Hydrolysis of lignocellulosic biomass for bioethanol production. In: Pandey, A., Larroche, C., & Gnansounou, E., (eds.), *Biofuels* (pp. 229–250). Academic Press: United Kingdom.
33. Saritha, M., Arora, A., & Lata, (2012). Biological pretreatment of lignocellulosic substrates for enhanced delignification and enzymatic digestibility. *Indian J. Microbiol., 52,* 122–130.
34. Voet, D., & Voet, J. G., (2006). *Bioquímica.* Médica Panamericana.
35. Castellanos, O., Ramírez, D. C., & Montañez, V. M., (2006). Perspectives in developing industrial enzymes by using technological intelligence. *Ing. Investig., 26,* 52–67.
36. Sarrouh, B., Santos, T. M., Miyoshi, A., Dias, R., & Azevedo, V., (2012). Up-to-date insight on industrial enzymes applications and global market. *J. Bioprocess. Biotech., 4,* 2.
37. Ortiz, M. M. L., & Vélez, D. U., (2010). Determinación de la actividad lignocelulolítica en sustrato natural de aislamientos fúngicos obtenidos de sabana de pastoreo y de bosque secundario de sabana inundable tropical. *Cienc. del Suelo., 28,* 169–180.
38. Chacón, O., & Waliszewski, K. N., (2005). Commercial cellulases preparations and their applications in extractives processes. *Univ. y Cienc., 21,* 111–120.
39. López Domínguez, C. M. (2014). Evaluation in Citric Wastes Saccharification Employing Cellulosomes Producers Microorganisms [Master's thesis dissertation,

Center for Research and Assistance in Technology and Design of the Jalisco State, A. C.]. *CIATEJ Repository.* https://ciatej.repositorioinstitucional.mx/jspui/handle/1023/53.

40. Volynets, B., & Dahman, Y., (2011). Assessment of pretreatments and enzymatic hydrolysis of wheat straw as a sugar source for bioprocess industry. *Int. J. Energy Environ., 2*, 427–446.

41. Cuervo, L., Folch, J. L., & Quiroz, R. E., (2009). Lignocelulosa como fuente de azúcares para la producción de etanol. *BioTecnología, 13*, 11–25.

42. Shaikh, H. M., Adsul, M. G., Gokhale, D. V., & Varma, A. J., (2011). enhanced enzymatic hydrolysis of cellulose by partial modification of its chemical structure. *Carbohydr. Polym., 86*, 962–968.

43. Janusz, G., Kucharzyk, K. H., Pawlik, A., Staszczak, M., & Paszczynski, A. J., (2013). Fungal laccase, manganese peroxidase and lignin peroxidase: Gene expression and regulation. *Enzyme Microb. Technol., 52*, 1–12.

44. Hofrichter, M., (2002). Lignin conversion by manganese peroxidase (MnP). *Enzyme Microb. Technol., 30*, 454–466.

45. Fujii, K., Uemura, M., Hayakawa, C., Funakawa, S., & Kosaki, T., (2013). Environmental control of lignin peroxidase, manganese peroxidase, and laccase activities in forest floor layers in humid Asia. *Soil Biol. Biochem., 57*, 109–115.

46. Manoj, K. M., & Hager, L. P., (2008). Chloroperoxidase, a Janus enzyme. *Biochemistry, 47*, 2997–3003.

47. Zhao, M., Zeng, Z., Zeng, G., Huang, D., Feng, C., Lai, C., Huang, C., et al., (2012). Effects of ratio of manganese peroxidase to lignin peroxidase on transfer of ligninolytic enzymes in different composting substrates. *Biochem. Eng. J., 67*, 132–139.

48. Bribiesca, B. L. C., (2013). Enzimas xilanolíticas bacterianas y sus aplicaciones industriales. *Rev. Espec. en Ciencias la Salud., 16*, 19–22.

49. Rasmussen, H., Sørensen, H. R., & Meyer, A. S., (2014). Formation of degradation compounds from lignocellulosic biomass in the bio-refinery: Sugar reaction mechanisms. *Carbohydr. Res., 385*, 45–57.

50. Cola, P., Procópio, D. P., Alves, A. T. D. C., Carnevalli, L. R., Sampaio, I. V., Da Costa, B. L. V., & Basso, T. O., (2020). Differential effects of major inhibitory compounds from sugarcane-based lignocellulosic hydrolysates on the physiology of yeast strains and lactic acid bacteria. *Biotechnol. Lett., 42*, 571–582.

51. Papagianni, M., (2004). Fungal Morphology and metabolite production in submerged mycelial processes. *Biotechnol. Adv., 22*, 189–259.

52. Subramaniyam, R., & Vimala, R., (2012). Solid state and submerged fermentation for the production of bioactive substances: A comparative study. *Int. J. Sci. Nat., 3*, 480–486.

53. Hansen, G. H., Lübeck, M., Frisvad, J. C., & Lübeck, P. S., (2015). Production of cellulolytic enzymes from ascomycetes: Comparison of solid state and submerged fermentation. *Process Biochem., 50*, 1327–1341.

54. Aidoo, K. E., Hendry, R., & Wood, B. J. B., (1982). Solid substrate fermentations. In: *Advances in Applied Microbiology* (Vol. 28, pp. 201–237).

55. Hesseltine, C. W., (1972). Biotechnology report: Solid state fermentations. *Biotechnol. Bioeng., 14*, 517–532.

56. Behera, S. S., Ray, R. C., & Das, U., (2019). Microorganisms in fermentation. In: Berenjian, A., (ed.), *Essentials in Fermentation Technology. Learning Materials in Biosciences* (pp. 1–39). Springer.

57. Singhania, R. R., Sukumaran, R. K., Patel, A. K., Larroche, C., & Pandey, A., (2010). Advancement and comparative profiles in the production technologies using solid-state and submerged fermentation for microbial cellulases. *Enzyme Microb. Technol., 46*, 541–549.

58. Kumar, P. K. R., & Lonsane, B. K., (1989). Microbial production of gibberellins: State of the art. *Adv. Appl. Microbiol., 34*, 129–139.

59. Venkata, R. P., Jayaraman, K., & Lakshmanan, C. M., (1993). Production of lipase by *Candida rugosa* in solid state fermentation: Determination of significant process variables. *Process Biochem., 28*, 385–389.

60. Manpreet, S., Sawraj, S., Sachin, D., Pankaj, S., & Banerjee, U. C., (2005). Influence of process parameters on the production of metabolites in solid-state fermentation. *Malaysian Journal Microbiol., 1*, 1–9.

61. Pérez-Guerra, N., Torrado-Agrasar, A., López-Macias, C., & Pastrana, L., (2003). Main characteristics and applications of solid substrate fermentation. *Electron. J. Environ. Agric. Food Chem., 2*, 3.

62. Beuchat, L. R., (1983). Influence of water activity on growth, metabolic activities and survival of yeasts and molds. *J. Food Prot., 46*, 135–141.

63. Hahn-Hägerdal, B., (1986). Water activity: A possible external regulator in biotechnical processes. *Enzyme Microb. Technol., 8*, 322–327.

64. Mattiasson, B., & Hahn-Hägerdal, B., (1982). Microenvironmental effects on metabolic behavior of immobilized cells a hypothesis. *Eur. J. Appl. Microbiol. Biotechnol., 16*, 52–55.

65. Chirife, J., Favetto, G., & Scorza, O. C., (1982). The water activity of common liquid bacteriological media. *J. Appl. Bacteriol., 53*, 219–222.

66. Gervais, P., Bensoussan, M., & Grajek, W., (1988). Water activity and water content: Comparative effects on the growth of *Penicillium roqueforti* on solid substrate. *Appl. Microbiol. Biotechnol., 27*, 389–392.

67. Troller, J. A., (1983). Influence of water activity on microorganisms in foods. *Food Technol., 34*, 76–82.

68. Larroche, C., Besson, I., Dussap, C. G., Bourrust, F., & Gros, J. B., (1993). Characterization of water distribution in cell pellets using nonlabelled sodium thiosulfate as an interstitial space marker. *Biotechnol. Prog., 9*, 214–217.

69. Sargantanis, J., Karim, M. N., Murphy, V. G., Ryoo, D., & Tengerdy, R. P., (1993). Effect of operating conditions on solid substrate fermentation. *Biotechnol. Bioeng., 42*, 149–158.

70. Ramesh, M. V., & Lonsane, B. K., (1990). Critical importance of moisture content of the medium in alpha-amylase production by *Bacillus licheniformis* M27 in a solid-state fermentation system. *Appl. Microbiol. Biotechnol., 33*, 501–505.

71. Sharma, K. M., Kumar, R., Panwar, S., & Kumar, A., (2017). Microbial alkaline proteases: Optimization of production parameters and their properties. *J. Genet. Eng. Biotechnol., 15*, 115–126.

72. Dos, S. A. J. G., & Sato, H. H., (2018). Microbial proteases: Production and application in obtaining protein hydrolysates. *Food Res. Int., 103*, 253–262.

73. Elias, M., Wieczorek, G., Rosenne, S., & Tawfik, D. S., (2014). The universality of enzymatic rate-temperature dependency. *Trends Biochem. Sci., 39*, 1–7.

74. Hashemi, M., Shojaosadati, S. A., Razavi, S. H., Mousavi, S. M., Khajeh, K., & Safari, M., (2012). The efficiency of temperature-shift strategy to improve the production of α-amylase by *Bacillus* Sp. in a solid-state fermentation system. *Food Bioprocess Technol., 5*, 1093–1099.

75. Cannel, E., & Moo-Young, M., (1980). Solid-state fermentation systems. *Process Biochem., 15*, 24—28.

76. Sandhya, C., Nampoothiri, K. M., & Pandey, A., (2005). Microbial proteases. In: Barredo, J. L., (ed.), *Microbial Enzymes and Biotransformations* (pp. 165–179). Springer.

77. Kumar, C. G., & Takagi, H., (1999). Microbial alkaline proteases. *Biotechnol. Adv., 17*, 561–594.

78. Kasana, R. C., Salwan, R., & Yadav, S. K., (2011). Microbial proteases: Detection, production, and genetic improvement. *Crit. Rev. Microbiol., 37*, 262–276.

79. Yegin, S., Fernández, L. M., Guvenc, U., & Goksungur, Y., (2010). Production of extracellular aspartic protease in submerged fermentation with *Mucor mucedo* DSM 809. *African J. Biotechnol., 9*, 6380–6386.

80. Çalık, P., Bilir, E., Çalık, G., & Özdamar, T. H., (2002). Influence of pH conditions on metabolic regulations in serine alkaline protease production by *Bacillus licheniformis*. *Enzyme Microb. Technol., 31*, 685–697.

81. Peñalva, M. A., & Arst, H. N. J., (2002). Regulation of gene expression by ambient pH in filamentous fungi and yeasts regulation of gene expression by ambient pH in filamentous fungi and yeasts. *Microbiol. Mol. Biol. Rev., 66*, 426–446.

82. Hesse, S. J. A., Ruijter, G. J. G., Dijkema, C., & Visser, J., (2002). Intracellular pH homeostasis in the filamentous fungus *Aspergillus niger*. *Eur. J. Biochem., 269*, 3485–3494.

83. Singh, S. K., Tripathi, V. R., Jain, R. K., Vikram, S., & Garg, S. K., (2010). An antibiotic, heavy metal resistant and halotolerant *Bacillus cereus* SIU1 and its thermoalkaline protease. *Microb. Cell Fact., 9*, 59.

84. Wang, L., Ridgway, D., Gu, T., & Moo-Young, M., (2005). Bioprocessing strategies to improve heterologous protein production in filamentous fungal fermentations. *Biotechnol. Adv., 23*, 115–129.

85. Lonsane, B. K., Ghildyal, N. P., Budiatman, S., & Ramakrishna, S. V., (1985). Engineering aspects of solid state fermentation. *Enzyme Microb. Technol., 7*, 258–265.

86. Lean, M. E. J., (2019). Principles of human nutrition key points. *Medicine (Baltimore), 47*(3), 140–144.

87. WHO, (2002). *Joint FAO/WHO/UNU Expert Consultation on Protein and Amino Acid Requirements in Human Nutrition*. Geneve, Switzerland.

88. Fasolin, L. H., Pereira, R. N., Pinheiro, A. C., Martins, J. T., Andrade, C. C. P., & Ramos, O. L., (2019). Emergent food proteins - towards sustainability, health and innovation. *Food Res. International, 125*, 108586.

89. Villalobos, A. C., (2004). Perspectivas actuales de la proteína unicelular (SCP) en la agricultura y la industria. *Agron. Mesoam.*, 93–106.

90. Ritala, A., Toivari, M., & Wiebe, M. G., (2017). Single cell protein — state-of-the-art, industrial landscape and patents 2001–2016. *Front Microbiol., 8*.

91. León, C., (2005). *Influencia de la concentración de melaza de Saccharum officinarum L. "caña de azúcar" en la producción de proteína unicelular de Candida utilis var. major*. MSc dissertarion, Universidad Nacional de Trujillo, Trujillo, Perú.

92. Crueger, W., & Crueger, A., (1993). *Biotechnology: Manual of Industrial Microbiology*. Acribia Publisher. ISBN: 84-200-0743-9.

93. Suharto, I., (1983). Mini-fermentation technology to produce single-cell protein from molasses [as Animal Feeds in Indonesia]. *Food Nutr. Bull. Suppl., 7*.

94. Raimbault, M., Soccol, C. R., & Chuzel, G., (1998). In: D'Encre, S. L. G., (ed.), *International Training Course on Solid State Fermentation*. Montpellier.

95. Bajpai, P., (2017). Single-cell protein from lignocellulosic wastes. In: *Single Cell Protein Production from Lignocellulosic Biomass* (pp. 41–58). Springer: Singapore.

96. Reihani, S. F. S., & Khosravi-darani, K., (2019). Influencing factors on single-cell protein production by submerged fermentation: A review. *Electron. J. Biotechnol. Rev., 37*, 34–40.

97. Agri-Food and Fisheries Information Service [SIAP] (January 9, 2017). Agricultural Production. México. https://www.gob.mx/siap/acciones-y-programas/produccion-agricola-33119.

98. Auria, R., Morales, M., Villegas E., & Revvah S. (1993). Influence of mold growth on the pressure drop in aerated solid state fermentors. *Biotechnol. Bioeng., 41*, 1007–1113.

99. Biesebeke, R., Ruijter, G., Rahardjo, Y., Hoogschagen, M., Heerikhuisen, M., Levin, A., van Driel, K., Schutyser, M., Dijksterhuis, J., Zhu, Y., Weber, F., De Vos, W., Hondel, K., Rinzema, A., & Punt, P. (2002). *Aspergillus oryzae* in solid-state and submerged fermentations: Progress report on a multi-disciplinary project. *FEMS Yeast. Res., 2*, 245–248.

100. Soccol, C. R., Costa, E. S., Letti, L. A., Karp, S. G., Woiciechowski, A., & Vandenberghe, L. P. (2017). Recent developments and innovations in solid state fermentation. *Biotechnol. Res. Innov., 1*, 52–71.

CHAPTER 14

Impact of Functional Ingredients from Plant Food Byproducts on Human Gut Microbiota

RICARDO GÓMEZ-GARCÍA, DÉBORA A. CAMPOS, ANA R. MADUREIRA, and MANUELA PINTADO

Universidade Católica Portuguesa, CBQF-Centro de Biotecnologia e Química Fina-Laboratório Associado, Escola Superior de Biotecnologia, Rua Diogo Botelho 1327, Porto – 4169-005, Portugal, E-mail: mpintado@porto.ucp.pt (M. Pintado)

ABSTRACT

Plant food byproducts (PFBs) from food processing industries have been over produced and mismanaged worldwide causing some concerns related with food losses, economics, and environmental issues. These byproducts include diverse vegetable tissues among peels, stems, leaves, kernels, and pomace, which have been accepted as innovative, low-cost, and natural sources of bioactive molecules, such as polyphenols, proteins, polysaccharides, and dietary fiber (DF) with functional properties. Thus, based on these value-added characteristics some biotechnological studies have highlighted the importance to re-use and re-cycle the PFBs for the development of novel functional ingredients and foods. Nowadays, functional ingredients are well-recognized by their positive effect exhibited on human health due to the high content of DF and polyphenols, which have been shown positive modulation on human intestinal microbiota mainly promoting specific bacteria groups such as *Firmicutes, Lactobacillus, and Bifidobacterium*, usually associated with the prevention of several diseases. Hence, this chapter describes relevant information regarding the employment of PFBs as functional ingredients and their principal interactions for a positive modulation of gut microbiota.

14.1 INTRODUCTION

Nowadays, the food byproducts production is assessed to be around 1.3 billion tons annually, such huge production is primarily generated by the food producers and processing industries with an estimated economic cost of 990 billion dollars. In Europe, around 96 million metric tons of vegetables were produced, corresponding to 8.5% of the global production and circa 30% of such vegetables have been rejected as byproducts and poorly managed [1, 2]. Thus, there exist social, economic, and environmental concerns to improve profitable and valorization of these vegetable byproducts. On the other hand, the increasing claim of natural, safe, and clean label food products, with beneficial and health-promoting characteristics, has contributed to diverse research studies for the development of novel functional food ingredients through integral valorization of some PFBs within the framework of circular economy and environmental pollution prevention [3, 4]. Furthermore, the PFBs have been well-categorized as rich sources of different high value-added compounds, such as dietary fiber (DF) and bioactive compounds. In turns, DF is divided in insoluble dietary fiber (IDF), which includes cellulose, hemicellulose, and lignin, and soluble dietary fiber (SDF), which includes non-starch polysaccharides, β-glucans, gums, pectin, oligosaccharides, or inulin. The SDF has shown several technological and functional properties such as water-holding, swelling capacity, gel formation and increasing viscosity [5]. Additionally, bioactive compounds such as polyphenols are classified into diverse groups, such as tannins, lignans, flavonoids, and phenolic acids, which have been extensively reported as beneficial natural molecules for human health and wellbeing [6]. The combination between DF and polyphenols has been shown an important role to modulate the composition and metabolic functions of the microbial communities from human gut microbiota, helping to prevent or reduce the risk of certain chronic illnesses like cancers, diabetes, cardiovascular, and inflammatory disorders [7, 8]. Hence, the DFs associated with polyphenols create a multifunctional product (combining both biological effects), enhancing the bioactive and functional properties. Hence, by virtue of these functional/beneficial properties, and based on their richness in DF and polyphenols, the PFBs could be used as functional ingredients to develop new food products. The specific aims of this chapter are to gather and increase research interest in the valorization of the PFBs, describing their positive contribution on gut microbiota and consequently on health. This will further encourage novelty in the design and development of functional products that can solve problems related to food losses, waste, and pollution, focusing on their potential applications in the modern food industries.

14.2   GUT MICROBIOTA AND ROLE ON HEALTH

The gut microbiota is the surviving microbial population present in the colon, such population is varied and ample (10^{14} cells) and comprises of different microorganism as bacteria, archaea, and eukaryotes that play an important role with the host. *Firmicutes* and *Bacteroidetes* are the major bacteria found in the human gut and in a minor amount are the *Proteobacteria* and *Actinobacteria*. Together, all these strains constitute 93–98% of gut microbiota [9]. *Bifidobacterium, Lactobacillus, Bacteroides, Clostridium, Escherichia, Streptococcus, and Ruminococcus* are the most usual bacteria living in the gut. These bacteria are vital for the health and wellbeing of their host, having a helpful function in the reduction of disease or in the maintenance of the immunological activity, energy consumption and even in the brain activity. A healthy and strong structure of gut microbiota can build a physical barrier against infections, whereas disturbance in the balance of gut ecology (dysbiosis) causes higher susceptibility to pathogens. Diverse reports have discussed the relationship between dysbiosis and disease development, involving obesity, inflammatory bowel disease (IBD) and cancer [10]. Obesity and type II diabetes are diseases states associated with gut microbiota composition and composition changes. In addition, dysbiosis is also linked with anxiety and depression due to the consumption of probiotics produces anxiolytic- and antidepressant-like activity in animal reports. On the other hand, food has a fundamental role in influencing the composition and the activity of the heterogeneous microbial population in the gut, providing nutrients such as carbohydrates, proteins, and fats, as well as, is essential for the obtainment of energy from food and even the conservation of the immune system. A rich diet of fibers that may promote gut microbiota modulation and increased short chain fatty acids (SCFA) concentrations generated by microbiota are generally considered to provide gastrointestinal health benefits. An increasing of SCFA production within gut microbiota reduces luminal pH, which help to prevent colonization and infections from pathogenic bacteria [11].

14.3 MODULATION OF GUT MICROBIOTA

Currently, the gut microbiota is considered a metabolic organ that can modulate nutrient absorption and interact with the immune system. Recent studies indicated that the valuation of the well-being includes the conservation and modulation of the gut microbiota, which is altered in its composition and function by diet. Diet is one of the most relevant factors that can modulate the gut microbiota, having a large range of effects on the host. Additionally, diet has been recognized as fundamental contributor to the composition and role of the human gastrointestinal microbiota. There are remarkable scientific data showing the changes in the abundance of gut microbes through consumption of DFs and prebiotics, some of such microbes are able to produce SCFA, which are linked to a range of health benefits [11]. Moreover, food diet containing many of different types of plants have been linked to greater gut bacteria diversity and some evaluations of certain fruits, vegetables, nuts, and seeds have significative impact in the gut microbiota and human health. This impact is derived from their content of fibers that have varied physicochemical properties, including solubility and fermentability, which influence bacterial fermentation profiles in the gastrointestinal tract. Soluble fibers, such as inulin, resistant maltodextrins, starch, and polydextrose are quickly fermented by gastrointestinal microbiota. Microbes from the human gut possess a range of carbohydrate-bindings modules and an extensive set of enzymes, commonly glycoside hydrolases, glycosyltransferases, polysaccharides lyases and carbohydrate esterases, enzymes that promote the hydrolysis of wide variety of food-based plants that contain fiber like fructans, pectins, β-glucans and arabinoxylans and resistant starches. These fibers are composed by a wide range of monosaccharides units linked by α- and β-bonds configuration, which may be more supportive of a varied microbial community than a diet that has less diverse substrates (refined diet) [12].

14.3.1 PREBIOTICS

Prebiotic compounds have been recognized for their ability to effect host microbiota to promote benefits in the host, for example, fructans (fructooligosaccharides (FOS) and inulin) and galactans (galactooligosaccharides (GOS)), acting through enrichment of *Lactobacillus* and *Bifidobacterium* spp. [13]. However, due to the developments and tools for microbiota studies, now the prebiotic concept has extended, such developments have improved the knowledge of the composition of the microbiota and enabled the identification of additional substances prompting colonization, as well as, by the comprehension that a wide array of beneficial microorganisms are disturbed by prebiotics and also, they could be effective at extraintestinal places directly or indirectly [14]. Currently, prebiotics are defined as a substrate (non-digestible ingredient) that is selectively employed by host microorganisms deliberating health benefits. On the other hand, various soluble fermentable fibers and other kind of DFs can be prebiotics because they are selectively employed by the host microbiota and promote health [15].

14.3.2 DIETARY FIBERS (DFS)

Dietary fiber (DF) can be classified in two groups such as water-soluble (pectin) and water-insoluble (cellulose and lignin). Soluble fiber has some favorable physiological functions, among the increment of the viscosity of food digesta, changing the rate of nutrient release and absorption in gastrointestinal tract and lowers blood cholesterol concentration. IDF has less physiological effects in the upper gastrointestinal tract and is fermented in less extent by the colonic microflora. However, insoluble DF plays a key role in intestinal regulation through mechanical peristalsis [1]. The degree of polymerization and the solubility of complex carbohydrates influences the site of fermentation during the transit in the human gastrointestinal tract. Soluble fiber, such as FOS and pectin are hydrolyzed by bacteria nearly in the gastrointestinal tract (the ileum and ascending colon), while cellulose (fibers less soluble), can be moderately fermented in the distal colon where the passage time is slower, and bacteria densities are higher. Therefore, fibers behave as substrates that are differentially metabolized by the gut microbes, generating SCFA. Polydextrose and soluble fiber, which are constituted by glucose monomers linked by $\alpha-$ and β-(1,2), α- and β-(1,3), α- and β-(1,4), and $\alpha-$ and β-(1-6)-bonds, can be used as substrate of certain gut microbes. Inulin-type fibers, which principally consist in a linear fructosyl polymer linked by β-(1,2) bonds attached to a terminal glucosyl residue by an α-(1,2) bonds provide selective stimulation for bifidobacterial [11].

14.3.3 SHORT CHAINS FATTY ACIDS (SCFA)

Microbial hydrolysis of complex polysaccharides to monosaccharides includes different biochemical routes and is carried out by the enzymatic production and activity. For example, Acetate, propionate, butyrate, and gas (H_2 and CO_2) are the main bacteria fermentative end SCFA products from the complex carbohydrates [16]. The production of SCFA indicate a fermentative process by the colon bacteria. The alteration of the SCFA concentration is manifested during the length of the gastro-intestinal tract, with the highest concentration near to the colon, which is the zone of the gastrointestinal tract with the biggest concentration of microbes and the lessening concentrations in the distal colon [17]. In turn, *Faecalibacterium prausnitzii* (butyrate-producing bacteria) and *Akkermansia muciniphila* (a mucin degrading bacterium) have both been linked with helpful health outcomes, including reduction of inflamma-tion and enhanced gut barrier role [18]. Butyrate is the key energy source of colonocytes and enterocytes. Propionate can be employed through conversion into glucose by intestinal gluconeogenesis or diffuse into the portal vein to be consumed as substrate for hepatic gluconeogenesis [19]. The SCFA are absorbed in the gut or used by the microbial popula-tion. However, a small concentration of SCFA, especially propionate and acetate, are found in peripheral circulation. Regarding acetate, the most prominent SCFA has been shown to cross the blood-brain barrier. Furthermore, despite acetate be an essential co-factor/metabolite for the growth of bacteria, it provides the capacity of bifidobacteria to inhibit enteropathogens and reduce the appetite of the host through the interac-tion with the central nervous system [20].

14.3.4 POLYPHENOLS

High percentage of polyphenols (95%) cannot be absorbed in the small intestine, they are metabolized in the colon by the residing bacteria, leading to the production of metabolites, which together with polyphenols can modulate the bacterial population in the gut. This phenomenon is attributed to their complex structure and high molecular weights of polyphenols. Biotransformation of polyphenols in the gut has been shown to be carried out by intestinal bacteria, for example gut bacteria deconjugate anthocyanins by cleaving their glycosidic bonds to form lower molecular weights compounds, and also some reports demonstrated the catabolism of quercetin, rutin, chlorogenic acid and caffeic acid to generate smaller phenolic acids by gut microbiota [8]. Moreover, polyphenols can alter the composition of microbial bacteria in the gut, due to their ability to inhibit growth of specific bacteria, especially pathogenic, such as *E. coli, S. aureus* and *C. albicans* [21]. In addition to act as antimicrobial agent against pathogens, polyphenols are also suggested to present their protective effect as prebiotics because of their inhibitory effect against pathogens, while beneficial bacteria are stimulated and are not affected by polyphenols. For example, epicatechin, and catechin not only inhibited foodborne pathogens, but also increased the count of probiotics like *Lactobacillus, Bifidobacterium spp.* and *Eubacterium rectale-C. coccoides* [16].

14.4 FUNCTIONAL INGREDIENTS FROM PLANT FOOD BYPRODUCTS (PFBS)

The PFBs are rich in fibers and polyphenols that have been contributing to bowel increasing health, weigh management, lower blood cholesterol levels and improve control of glycemic and insulin responses. Due to the composition and beneficial characteristics of PFBs, they have been employed for the development of functional ingredients [22]. Nowadays, functional foods are described as food ingredients that have diverse health benefits, preventing or reducing the risk of several illness, such as cancers, cardiovascular, and inflammatory disorders, and diabetes [23]. These natural materials could be used directly as a source of DF or as feedstock to obtain some oligosaccharides. These natural polymers have been used for the modulation of intestinal microbiota, such as *Bifidobacterium*, which have been associated to prevent colon, stomach, prostate, and breast cancer [6]. For example, corncobs byproducts were employed to produce xylo-oligosaccharides, non-digestible food ingredients with prebiotic effect, such ingredients were selectively used for *Bifidobacterium* and *Lactoba-cillus,* improving their growth due to their ability to produce hydrolytic enzymes (xylanase and β-xylosidase). Therefore, such bacteria help in the digestion of non-digestible oligosaccharides in the gut region, which are normally not digested in the upper gastrointestinal tract and produce SCFA resulting in the reduction of pH in the environment. The decrease in the pH has been well-documented that creates an acidic environment, which in turn decrease the number of pathogenic bacteria and permits the normal microflora to survive in the human intestine maintaining one's health [24]. On the other hand, polyphenols have been demonstrated to modulate gut microbiota, as well as, exhibited good inhibition properties against digestive enzymes (such as lipase, glucosidase, and amylase), minimizing hyperglycemia which is strongly associated with diabetes and obesity disorders [16, 25]. By virtue of these functional properties on human health and on the basis of their richness in fibers and polyphenols, PFBs could be used as functional ingredients to develop novel food products such as cookies, fiber bars or cereals, contributing to the increasing demand of healthy foods.

14.5   STABILITY THROUGHOUT GASTROINTESTINAL TRACT

Consumption of food rich in polyphenols does not guarantee their bioavailability during the digestive tract, thus their biological effect is not exerted. These compounds vary in their composition and have different stability behaviors due to their availability within the food matrix as well as when are subjected to the gastrointestinal tract environments. Diverse of these compounds have low bioavailability and are weakly absorbed due to their large molecular weight and low lipid solubility, thus they cannot be absorbed by passive diffusion [26]. The adsorption in the gut in many of the times varies depending on food composition, processing conditions or presence of other compounds [28]. However, there are some specific factors that could affect the adsorption of the molecules in the gut such as food structure and molecular weight of the compounds and chemical interactions. For example, polyphenols extracts showed bioavailability loss when were exposed to *in vitro* simulation digestion after mouth process, decreasing their antioxidant power when reach to the stomach and therefore, minimizing their health promotion activity in the gut. This effect could be explained by the strong affinity of phenolic compounds with proteins, which usually are present in human saliva, forming non-covalent and covalent linkages [29]. In addition, bioavailability is expressed as the amount of a nutrient that is digested, absorbed, and utilized in normal metabolism and bioaccessibility is a term used to describe the amount of an ingested nutrient that is available for absorption in the gut after digestion [30]. On the other hand, it has been reported that polyphenols can be released from food matrix in the upper area of gastrointestinal tract by direct solubilization in the intestinal liquids at physiological conditions (37°C, pH 1–7.5) and by the hydrolysis of digestive enzymes. Humans' enzymes are capable to hydrolyze only a scarce glycosidic bonds present in carbohydrates, including starch polysaccharides, by the action of pancreatic and amylase, and the disaccharides sucrose and lactose by disaccharidases, sucrase, and lactase [31]. Therefore, phenolic compounds embedded within DF could arrive at the colon and exert their biological functions, contributing to the intestinal health. Furthermore, many research groups have been focused on the formulation of food ingredients from PFBs, investigating the effect and stability of their functionality through *in vitro* simulators, which include controlled incubations in real or simulated gastric juices as well as gastrointestinal models that simulate the environment in the gastrointestinal transit, which can help in the understanding of the changes of an acidic environment (pH 2.0–2.5) in contact with pepsin, followed by the exposition to bile salts (0.3%, w/v, at pH 5) for different time periods and then observe the final products that could reach to the gut and exert their beneficial properties [32].

14.6   IMPACT OF PLANT FOOD BYPRODUCTS (PFBS) ON GUT MICROBIOTA

Currently, diverse experts in gut microbiota are expanded beyond foods and prebiotics to the dietary patterns, eating habits and food formulation methods, in addition to the development of functional ingredients from PFBs that could have positive impacts on gut microbiota. In this context, DFs and polyphenols have been studied for the formulation of certain food ingredients, owning to their functional and bioactive properties as well as by their effect on the bioavailability of bioactive compounds, since polyphenols are usually associated with non-starch polysaccharides, which during the upper intestine digestion of foods, a partial portion of bioactive compounds pass to the blood system, whereas the remaining part reaches the colon and is fermented by the gut microbiota [33]. Furthermore, researchers have reported the prebiotic and bifidogenic positive effects exhibited by arabinoxylo-oligosaccharides (AXOS) from wheat bran through *in vitro* fermentation with human fecal microbiota where they showed stimulatory effect upon bifidobacterial population at the same order compared with pure FOS standard. Additionally, SCFA (acetate>propionate>butyrate) and lactate were produced at higher amounts than the controls. Remarkable information about the acetate-to-propionate ratios in gut have been proposed as a positive indicator of a hypolipidemic affect resulting from an inhibited biosynthesis of cholesterol and fatty acids in liver, which decrease blood lipid level [34]. Others have utilized pomegranate peel flour to evaluate the effect on polyphenols during *in vitro* gastrointestinal tract simulation and its ability to be fermented by colonic bacteria. The results showed 35.90 and 64% of bioaccessibility for phenolic and flavonoid compounds, respectively, which demonstrate the stability of these compounds, and their capacity to exert their health-related benefits, including potential cardiovascular benefits. On the other hand, pomegranate flour after digestion shown to be a suitable substrate for intestinal bacteria, producing beneficial SCFA, which modulate different processes including cell proliferation and differentiation, hormone secretion and activation of immune/inflammatory reactions [35]. In a recent study, functional flours (rich in DF and polyphenols) were developed from pineapple stems and peels where through the gastrointestinal tract simulation, a high phenolic content, comprised mainly by chlorogenic, caffeic, coumaric, and ferulic acids, was released from the pineapple flours. The antioxidant activity was conserved after digestion, which could suggest an antioxidant ecosystem within human gut. Also, after digestion, oligosaccharides (<1,000 Da) were detected, which were produced from the hydrolysis of the complex polysaccharides present in the pineapple matrix by the enzymatic activity. On the other hand, through fecal fermentation, the digested pineapple flours promoted the evolution of beneficial gut strains and caused the production of organic acids specially, acetic, propionic, and butyric acids. Peel flour exhibited a positive effect on *Lactobacillus* spp.

and *Bifidobacterium* spp., while for *Firmicutes* showed a negative effect. Moreover, stem flour demonstrated progressive effect on *Lactobacillus* spp., *Bifidobacterium* spp., *Firmicutes*, and *Bacteroidetes* (ca. 30% after 48 h) than peel flour and the positive control (FOS) [36].

14.6.1   *IN VITRO FERMENTABILITY ASSESSMENT*

Fecal fermentation is a laboratory level technique, which employs human or mice feces in anaerobic conditions to evaluate the modulation of resident microbes, but especially the stimulation of prebiotic effect of the target sample. Such technique begins, collecting the fresh samples (fecal feces) in sterile and anaerobic environments from healthy donors (at least 5 donors), who were free of any known metabolic and gastrointestinal diseases, zero ingestion of probiotic or prebiotic supplements or any antibiotics during the latest 3 months prior of fecal donation. Fecal inoculum is prepared in a reduced physiological salt solution (Table 14.1) at final concentration of 100 g feces/L at pH 6.8 [37]. Fermentation medium is composed by nutrients, which helps microorganisms' growth, adjusted at pH 6.8 [34]. The fermentation assessment is performed by mixing the target sample (2% w/v) and inoculum (2% v/v) in the medium at 37°C inside of an anaerobic chamber (5% H_2, 10% CO_2 and 85% N_2) and compared against positive (FOS) and negative (without sample) control, both prepared as described for the target sample. Additionally, for a better understanding of the microorganism's evolution and metabolic activities modulation during fermentation process, a sampling should be analyzed at different periods of time (e.g., 0, 12, 24, and 48 h). Cells are harvested from fermentation samples by centrifugation, the solid particles are stored at –20°C for their genomic DNA extraction and study, while the supernatants are prepared for analysis of pH, carbohydrates, organic acids, and phenolic content.

TABLE 14.1   Chemical Composition of Fecal and Fermentation Medium for *In Vitro* Fermentation Assay

	Concentration	
	g/L	% (v/v)
1. Fecal inoculum		
Reduced physiological salt solution (RPS: cysteine-HCl)	0.5	–
NaCl	8.5	–

TABLE 14.1 *(Continued)*

	Concentration	
2. Fermentation medium	–	–
Trypticase soya broth (TSB) without dextrose	5	–
Bactopeptone	5	–
Yeast nitrogen based	5	–
Cysteine-HCl	0.5	–
Trace mineral solution	–	1
Solution A	–	1
*NH_4Cl	100	–
*$MgCl_2 \cdot 6H_2O$	10	–
*$CaCl_2 \cdot 2H_2O$	10	–
Solution B	–	0.2
**$K_2HPO_4 \cdot 3H_2O$	200	–
***Resazurin	–	0.2

Note: The concentrations given were calculated for 1 L of distillate-sterile water.
*Reagents for solution A;
**Regents for solution B;
***Reagent is used from a solution of 0.5 g/L.

14.6.2 GUT MICROBIOME MODULATION

The alteration of food lifestyles has been shown modulation of the gut microbiome composition, which in turn can improve health or even disease avoidance. The gut microbiome covers the totally of microorganism living the gastrointestinal tract, which is mainly composed by bacteria with an estimated value of 100 trillion cells. The bacterial societies vary in structure along the digestive tract and adapt through life, according to lifestyle and nourishment of the host. Some reports have exhibited that the gut microbiome matches our human genome with about 100 times more DNA, which provide considerably to our physiology and metabolism and its composition variations with age, diet, among other factors. Most of these genes are involved in energy fabrication and metabolism, conferring humans the capacity to live on widely diverse diets [19]. Nowadays, several analyzes of fingerprinting have been developed for exploring the gut microbiota, which include temperature gradient gel electrophoresis (TGGE), denaturing gradient gel electrophoresis (DGGE), terminal restriction fragment length polymorphism (T-RFLP), quantitative PCR (qPCR) and fluorescence *in situ* hybridization (FISH), or plasmid-clone capillary Sanger sequencing. These methods are semiquantitative and bring a quickly outline of the microbiota but generally do not offer complete taxonomic data. Microarrays with analyzes corresponding to 16S rRNA sequences are high-throughput tools for describing abundance and diversity. Direct sequencing of 16S rRNA genes has been progressively employed for measuring microbial diversity and abundance in the human gut due to its low costs of sequencing, methods for data analysis, new bioinformatics algorithms and better databases with sequence of known taxonomy [38]. New invention on sequencing tools have allowed the identification of many cultivable and non-cultivable microorganism, such is the case of Illumina MiSeq, a high-performance molecular technology with better use for colonic microbiota identification, based on sequencing information from high resolution and strength. A recent study, employing mango peel and pulp to evaluate microbiota changes during colonic fermentation, showed that through sequencing on the Illumina MiSeq platform for 16 rRNA gene analysis was able to identify and quantify diverse microorganism genera, such as *Bifidobacterium, Buttiauxella, Faecalibacterium, Roseburia, Eubacterium, Fusicatenibacter, Holdemanella, Catenibacterium, Phascolarctobacterium, Collinsella, Prevotella, and Bacteroides* [14].

14.6.3   OTHER SIDE IMPACTS BY GUT MODULATION

Some research reports have remarked the impact of gut microbiota not only in the development of allergenic and intestinal diseases, but also in the modulation of depression, anxiety, and autoimmune syndromes. Nevertheless, an emerging field regarding gut microbiota has been rising, which will lead to a novel route of personalized medicine [39].

In 2010, the pharmacomicrobiomics designation was first presented and is define as the effect produced upon gut microorganism by a determined stimulus, in this case, a determined drug. Moreover, such term examines the effect of drugs variations within the human gut microbiome, how the gut microorganisms and their production of hydrolytic enzymes can modify the bioavailability, clinical efficacy and toxicity of some drugs and their mechanisms.

The studies on this field, have been mainly on predicting the effect of some treatments, primary in medicines applied to the treatment of cancer, but also, have been widely studied to promote human autoimmunity, which can potentially lead to application of pharmacomicrobiomics in precision medicine [40]. The application of pharmacomicrobiomics in human treatments would allow to have a diverse diagnostic perspective, since the patients would be tested for microbial species, genes, transcripts, and/or proteins that affect drug metabolism, small-molecule transport or immunoprotective responses, which will lead to a best development of therapeutic action on the basis of pretreatment gut microbial characteristics [41]. Several phases should be done before a complete implementation/ application, since there still exist a lack in the scientific knowledge on this matter, especially given the broad inferences of multiple sicknesses areas, multiple drugs, and conventional therapeutics [42].

14.7   CONCLUSIONS

Diet is a fundamental factor in influencing the abundances of human gut microbes and their production of metabolites, which can exert human health benefits. Therefore, the results obtained from the employment of PFBs as potential functional ingredients rich in DF and polyphenols have been shown a positive modulation on gut microbiota, especially on *Bifidobacterium* and *Lactobacillus* bacteria, which can generate health beneficial SCFA. Besides, the use of PFBs could help to reduce the cost and the environmental concerns associated with their disposal and poor management since these byproducts could be reincorporated to the food industrial chains focused on the development of functional foods that promotes health benefits.

ACKNOWLEDGMENTS

The authors would like to thank the National Council of Science and Technology (CONACyT, Mexico) for PhD fellowship support granted to Gómez-García Ricardo and also the authors would like to thank to the project Co-promoção n° 016403, "MULTIBIOREFINERY," supported by Programa Operacional Competitividade e Internacionalização e pelo Programa Operacional Regional de Lisboa, na sua componente FEDER, e pela Fundação para a Ciência e Tecnologia and project UID/Multi/50016/2013, administrated by FCT.

KEYWORDS

- **dietary fiber**
- **fructooligosaccharides**
- **functional ingredients**
- **gut microbiota**
- **polyphenols**
- **vegetable byproducts**

REFERENCES

1. Trigo, J. P., Alexandre, E. M. C., Saraiva, J. A., & Pintado, M. E., (2019). High value-added compounds from fruit and vegetable byproducts - characterization, bioactivities, and application in the development of novel food products. *Crit. Rev. Food Sci. Nutr.*, 1–29.
2. Campos, D. A., Gómez-García, R., Vilas-Boas, A. A., Madureira, A. R., & Pintado, M. M., (2020). Management of fruit industrial by-products—a case study on circular economy approach. *Molecules, 25*, 320.
3. Campos, D. A., Ribeiro, T. B., Teixeira, J. A., Pastrana, L., & Pintado, M. M., (2020). Integral valorization of pineapple (*Ananas comosus* L.) byproducts through a green chemistry approach towards added value ingredients. *Foods, 9*, 60.
4. Gómez-García, R., Campos, D. A., Aguilar, C. N., Madureira, A. R., & Pintado, M., (2020). Valorization of melon fruit (*Cucumis melo* L.) byproducts: Phytochemical and biofunctional properties with emphasis on recent trends and advances. *Trends Food Sci. Technol., 99*, 507–519.
5. Macagnan, F. T., Rodrigues, D. S. L., Sampaio, R. B., De Moura, F. A., Bizzani, M., & Picolli Da, S. L., (2015). Biological properties of apple pomace, orange bagasse and passion fruit peel as alternative sources of dietary fibre. *Bioact. Carbohydrates Diet. Fibre, 6*, 1–6.
6. Veiga, M., Costa, E. M., Silva, S., & Pintado, M., (2018). Impact of plant extracts upon human health : A review Impact of plant extracts upon human health : A review. *Crit. Rev. Food Sci. Nutr.*, 1–14.
7. Holscher, H. D., (2017). Dietary fiber and prebiotics and the gastrointestinal microbiota. *Gut Microbes, 8*, 172–184.

8. Lavefve, L., Howard, L. R., & Carbonero, F., (2020). Berry polyphenols metabolism and impact on human gut microbiota and health. *Food Funct., 11*, 45–65.

9. Mota De Carvalho, N., Costa, E. M., Silva, S., Pimentel, L., Fernandes, T. H., & Pintado, M. E., (2018). Fermented foods and beverages in human diet and their influence on gut microbiota and health. *Fermentation, 4*, 1–13.

10. Saffouri, G. B., Shields-Cutler, R. R., Chen, J., Yang, Y., Lekatz, H. R., Hale, V. L., Cho, J. M., et al., (2019). Small intestinal microbial dysbiosis underlies symptoms associated with functional gastrointestinal disorders. *Nat. Commun., 10*, 1–11.

11. Holscher, H. D., (2020). Diet affects the gastrointestinal microbiota and health. *J. Acad. Nutr. Diet., 120*, 495–499.

12. El Kaoutari, A., Armougom, F., Gordon, J. I., Raoult, D., & Henrissat, B., (2013). The abundance and variety of carbohydrate-active enzymes in the human gut microbiota. *Nat. Rev. Microbiol., 11*, 497–504.

13. Nava-Cruz, N. Y., Medina-Morales, M. A., Martínez, J. L., Rodríguez, R., & Aguilar, C. N., (2014). Agave biotechnology: An overview. *Crit. Rev. Biotechnol., 35*, 546–559.

14. Gutiérrez-Sarmiento, W., Sáyago-Ayerdi, S. G., Goñi, I., Gutiérrez-Miceli, F. A., Abud-Archila, M., Rejón-Orantes, J. C., Rincón-Rosales, R., et al., (2020). Changes in intestinal microbiota and predicted metabolic pathways during colonic fermentation of mango (*Mangifera indica* L.)-based bar indigestible fraction. *Nutrients, 12*, 683.

15. Gibson, K., Hutkins, G. R., Sanders, R., Prescott, M. E., Reimer, S. L., Salminen, R. A., & Verbeke, S. J., (2017). Expert consensus document: The international scientific association for probiotics and prebiotics (ISAPP) consensus statement on the definition and scope of prebiotics. *Nat. Rev. Gastroenterol. Hepatol., 14*, 491–502.

16. Kawabata, K., Yoshioka, Y., & Terao, J., (2019). Role of intestinal microbiota in the bioavailability and physiological functions of dietary polyphenols. *Molecules, 24*.

17. Aoe, S., Yamakata, C., Fuwa, M., Tamiya, T., Nakayama, T., Miyoshi, T., & Kitazono, E., (2019). Effects of BARLEYmax and high-β-glucan barley line on short-chain fatty acids production and microbiota from the cecum to the distal colon in rats. *PLoS One, 14*, e0218118.

18. Fu, X., Liu, Z., Zhu, C., Mou, H., & Kong, Q., (2019). Nondigestible carbohydrates, butyrate, and butyrate-producing bacteria. *Crit. Rev. Food Sci. Nutr., 59*, S130–S152.

19. Moco, S., Martin, F. P. J., & Rezzi, S., (2012). Metabolomics view on gut microbiome modulation by polyphenol-rich foods. *J. Proteome. Res., 11*, 4781–4790.

20. De Carvalho, N. M., Madureira, A. R., & Pintado, M. E., (2019). The potential of insects as food sources: A review. *Crit. Rev. Food Sci. Nutr.*, 1–11.

21. Pereira, A. P., Ferreira, I. C. F. R., Marcelino, F., Valentão, C., Andrade, P. B., Seabra, R., Estevinho, L., et al., (2007). Phenolic compounds and antimicrobial activity of olive (Olea europaea L. Cv. Cobrançosa) leaves. *Molecules, 12*, 1153–1162.

22. Lo, H. Y., Li, C. C., Chen, F. Y., Chen, J. C., Hsiang, C. Y., & Ho, T. Y., (2017). Gastro-resistant insulin receptor-binding peptide from *Momordica charantia* Improved the glucose tolerance in streptozotocin-induced diabetic mice via insulin receptor signaling pathway. *J. Agric. Food Chem., 65*, 9266–9274.

23. Silva, M. A., Albuquerque, T. G., Alves, R. C., Oliveira, M. B. P. P., & Costa, H. S., (2018). Melon (*Cucumis melo* L.) byproducts: Potential food ingredients for novel functional foods?. *Trends Food Sci. Technol., 98*, 181–189.

24. Chapla, D., Pandit, P., & Shah, A., (2012). Production of xylooligosaccharides from corncob xylan by fungal xylanase and their utilization by probiotics. *Bioresour. Technol., 115*, 215–221.

25. Sulaiman, S. F., & Ooi, K. L., (2014). Antioxidant and α-glucosidase inhibitory activities of 40 tropical juices from Malaysia and Identification of phenolics from the bioactive fruit juices of *Barringtonia racemosa* and *Phyllanthus acidus*. *J. Agric. Food Chem., 62*, 9576–9585.

26. Bordiga, M., Montella, R., Travaglia, F., Arlorio, M., & Coïsson, J. D., (2019). Characterization of polyphenolic and oligosaccharidic fractions extracted from grape seeds followed by the evaluation of prebiotic activity related to oligosaccharides. *Int. J. Food Sci. Technol., 54*, 1283–1291.

27. Campos, D. A., Madureira, A. R., Sarmento, B., Gomes, A. M., & Pintado, M. M., (2015). Stability of bioactive solid lipid nanoparticles loaded with herbal extracts when exposed to simulated gastrointestinal tract conditions. *Food Res. Int., 78*, 131–140.

28. Ribeiro, T. B., Oliveira, A., Campos, D., Nunes, J., Vicente, A. A., & Pintado, M., (2020). Simulated digestion of an olive pomace water-soluble ingredient: Relationship between the bioaccessibility of compounds and their potential health benefits. *Food Funct., 11*, 2238–2254.

29. Vilas-Boas, A. A., Oliveira, A., Jesus, D., Rodrigues, C., Figueira, C., Gomes, A., & Pintado, M., (2020). Chlorogenic acids composition and the impact of *in vitro gas*trointestinal digestion on espresso coffee from single-dose capsule. *Food Res. Int., 134*, 09223.

30. Quirós-Sauceda, A. E., Palafox-Carlos, H., Sáyago-Ayerdi, S. G., Ayala-Zavala, J. F., Bello-Perez, L. A., Álvarez-Parilla, E., De La Rosa, L. A., et al., (2014). Dietary fiber and phenolic compounds as functional ingredients: Interaction and possible effect after ingestion. *Food Funct., 5*, 1063–1072.

31. Blancas-Benitez, F. J., Mercado-Mercado, G., Quirós-Sauceda, A. E., Montalvo-González, E., González-Aguilar, G. A., & Sáyago-Ayerdi, S. G., (2015). Bioaccessibility of polyphenols associated with dietary fiber and *in vitro kine*tics release of polyphenols in Mexican 'Ataulfo' mango (*Mangifera indica* L.) byproducts. *Food Funct., 6*, 859–868.

32. Madureira, A. R., Amorim, M., Gomes, A. M., Pintado, M. E., & Malcata, F. X., (2011). Protective effect of whey cheese matrix on probiotic strains exposed to simulated gastrointestinal conditions. *Food Res. Int., 44*, 465–470.

33. Al-Sayed, H. M. A., & Ahmed, A. R., (2013). Utilization of watermelon rinds and sharlyn melon peels as a natural source of dietary fiber and antioxidants in cake. *Ann. Agric. Sci., 58*, 83–95.

34. Gullón, J. C., Gullón, B., Tavaria, P., Pintado, F., Gomes, M., Alonso, A. M., & Parajó, J. L., (2014). Structural features and assessment of prebiotic activity of refined arabino xylooligosaccharides from wheat bran. *J. Funct. Foods, 6*, 438–449.

35. Gullon, B., Pintado, M. E., Fernández-López, J., Pérez-Álvarez, J. A., & Viuda-Martos, M., (2015). *In vitro gas*trointestinal digestion of pomegranate peel (*Punica granatum*) flour obtained from co-products: Changes in the antioxidant potential and bioactive compounds stability. *J. Funct. Foods, 19*, 617–628.

36. Campos, D. A., Coscueta, E. R., Vilas-Boas, A. A., Silva, S., Teixeira, J. A., Pastrana, L. M., & Pintado, M. M., (2020). Impact of functional flours from pineapple byproducts on human intestinal microbiota. *J. Funct. Foods, 67*, 103830.
37. Madureira, A. R., Campos, D., Gullon, B., Marques, C., Rodríguez-Alcalá, L. M., Calhau, C., Alonso, J. L., et al., (2016). Fermentation of bioactive solid lipid nanoparticles by human gut microflora. *Food Funct., 7*, 516–529.
38. Karlsson, F., Tremaroli, V., Nielsen, J., & Bäckhed, F., (2013). Assessing the human gut microbiota in metabolic diseases. *Diabetes, 62*, 3341–3349.
39. Dinan, T. G., & Cryan, J. F., (2013). Melancholic microbes: A link between gut microbiota and depression?. *Neurogastroenterol. Motil., 25*, 713–719.
40. Scher, J. U., Nayak, R. R., Ubeda, C., Turnbaugh, P. J., & Abramson, S. B., (2020). Pharmacomicrobiomics in inflammatory arthritis: Gut microbiome as modulator of therapeutic response. *Nat. Rev. Rheumatol., 16*, 282–292.
41. Koppel, N., Rekdal, V. M., & Balskus, E. P., (2017). Chemical transformation of xenobiotics by the human gut microbiota. *Science, 356*, 1246–1257.
42. Lam, K. N., Alexander, M., & Turnbaugh, P. J., (2019). Precision medicine goes microscopic: Engineering the microbiome to improve drug outcomes. *Cell Host Microbe, 26*, 22–34.

CHAPTER 15

Trends, Analytical Approaches, and Applications of the VITEK System for Identification and Classification of Bacteria and Yeasts

ALAA KAREEM NIAMAH,[1] SHAYMA THYAB GDDOA AL-SAHLANY,[1]
DEEPAK KUMAR VERMA,[2,6] MAMTA THAKUR,[3]
BALARAM MOHAPATRA,[4] SMITA SINGH,[5]
MÓNICA L. CHÁVEZ-GONZÁLEZ,[6] CRISTÓBAL NOÉ AGUILAR,[6]
AMI R. PATEL,[7] and KOLAWOLE BANWO[8]

[1]*Department of Food Science, College of Agriculture, University of Basrah, Basra City, Iraq, E-mails: alaakareem2002@hotmail.com; alaa.niamah@uobasrah.edu.iq (A. K. Niamah)*

[2]*Department of Agricultural and Food Engineering, Indian Institute of Technology Kharagpur, Kharagpur – 721-302, West Bengal, India, E-mail: rajadkv@rediffmail.com*

[3]*Department of Food Engineering and Technology, Sant Longowal Institute of Engineering and Technology, Longowal – 148106, Punjab, India*

[4]*Department of Bioscience and Bioengineering, Indian Institute of Technology Bombay, Bombay, – 400076, Maharashtra, India*

[5]*Department of Life Sciences (Food Technology), Graphic Era (Deemed to be) University, Dehradun, Uttarakhand – 248002, India*

[6]*Bioprocesses and Bioproducts Research Group, Food Research Department, School of Chemistry, Autonomous University of Coahuila, Saltillo Unit – 25280, Coahuila, México, E-mail: cristobal.aguilar@uadec.edu.mx (C. N. Aguilar)*

[7]*Division of Dairy and Food Microbiology, Mansinhbhai Institute of Dairy and Food Technology-MIDFT, Dudhsagar Dairy Campus, Mehsana – 384-002, Gujarat, India*

[8]*Food Microbiology and Biotechnology Unit, Department of Microbiology, University of Ibadan, Oyo State, Nigeria*

ABSTRACT

The clinical and food testing microbiological laboratories required very rapid and careful methods for microbial identification and antibiotic sensitivity testing of microorganisms on regular basis. The VITEK II compact system is employed to identify microorganisms particularly bacteria, yeast, and molds automatically. The catalogs to identify VITEK system products are built using sets of numerous well-described bacterial and yeast strains estimated under several culture conditions. The technique of applied advanced colorimetry cards estimates the biochemical tests used in distinct microbial identification cards. VITEK II system database yields the identifications and classification of microorganisms through comparing the results to recognized microbial species-specific responses. in. VITEK II system has been successfully used to identify and classify numerous strains of bacteria including probiotics as well as pathogenic types in addition to yeasts and it had shown great accuracy in context to biochemical tests and genomic methods. The process of determining and classify the species of microorganisms is a stressful process and requires a long time. As compared to currently available methods, due to more rapid and accurate testing VITEK II can be considered as a very useful tool in clinical diagnostics where otherwise routine identification and testing of pathogenic strains is very risky and laborious task.

15.1   INTRODUCTION

Automation is growing at an early stage in microbiology laboratories than the automation rate in other laboratories, such as the hematological, clinical, and immunological laboratories [1–3]. In recent years, a range of automated methods has been developed for microorganisms' identification and antimicrobial sensitivity testing [4], which are based on automated interpreting scores of biochemical tests or on microdilution plates after 12 hours of incubation and photometric definition of microorganism growth. Development in modern technologies that can easily detect and classify bacteria and check antimicrobials now has both clinical and fiscal advantages [5–7]. Bacteria and yeasts are detected and studied for antimicrobial susceptibility in most general microbiology laboratories in the USA with automated systems. Utilization and price efficiency make them the preferred methods. The reference methods of culture broth microdilution and disk spreading assays are also used over the more laboratory general and clinical and laboratory requirements [8–10].

In 1970s, the VITEK device was introduced as an automated tool to identify the microorganisms and antimicrobial sensitivity. This system has been now a system called VITEK II which automatically proceeds with all the steps needed for identification of microorganisms and to check antimicrobial microorganisms after addition of the first inoculum [11]. The VITEK II is an automatic, growth-based microbiology method. In three designs (VITEK II compact, VITEK II and VITEK II XL), the system is obtained that differs in increasing sample numbers and in the automation process (Figure 15.1). These three devices have the same number of colorimetric reagent cards that are automatically incubated and clarified [12]. This device allows kinetic testing with a 15 min description of each test. The optical system allows the registration of turbidity, fluorescence, and colorimetric measurements with several channel colorimeter and Photometer readings. The VITEK method could provide reliable identification and antimicrobial sensitivity results with pure bacteria and yeast cultures, according to prior researchers [13, 14].

The goal of this chapter is to address trends in the 'VITEK II system' as an analytical methodology which include basic concepts, working principles, their different types and colorimetric reagent cards that is used to identify and classify bacteria and yeasts.

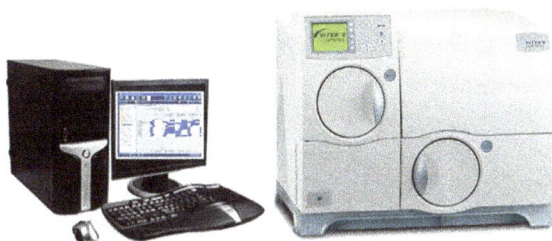

FIGURE 15.1   The VITEK II compact system [11, 15].

15.2   VITEK II: AN OVERVIEW

The main focus of this program is on the microbiological testing used in culture media, and microbiological laboratories for small to medium volumes of microbial production. The system has also provided an alternative for antimicrobial sensitivity testing by automated dilution and pipetting [16]. The colorimetric reagent cards of VITEK II system have four types based on microbes (Figure 15.2). Colorimetric reagent cards are inoculated by implanted vacuum machines with microbial isolate suspensions [17]. A microbial isolate suspension sample tube is transferred to the appropriate tablet (cassette) and the ID card is positioned in the adjacent region while the transfer pipe is inserted in the corresponding suspension tube. Up to 10–15 tests depend on the form of VITEK II device can be understood by the special tablet (cassette). The suspension of isolated microorganism is forced during the transfer pipeline to a micro tube to fill all the test wells after the vacuum cycle is used and air is re-inserted [18, 19].

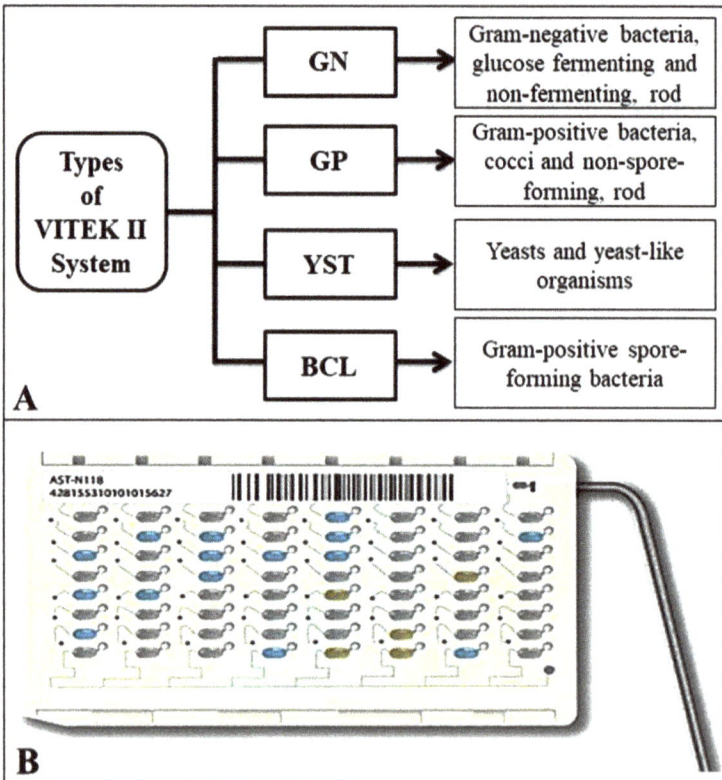

FIGURE 15.2   The VITEK II system. (A) Different four types of VITEK II system; and (B) the colorimetric reagent cards of VITEK II system [11, 15, 20].

A tool cuts the carrier tube and stamps the card before the carousel incubator is loaded into the VITEK II system to inoculate colorimetrical reagent cards [21]. The incubator carousel will hold between 30 and 60 cards. For the optic tool in the response readings, all colorimetric reagent type of card is incubated at 34.5–36 and 5C for 15 minutes. During the full incubation time, the data are taken at 15 minutes (Figure 15.3) [22, 23]. With the VITEK II compact, response period is decreased up to 22 hours for all bacteria species than the classic biochemical tests while the safety of biodata is maintained with overall acquiescence features [24, 25].

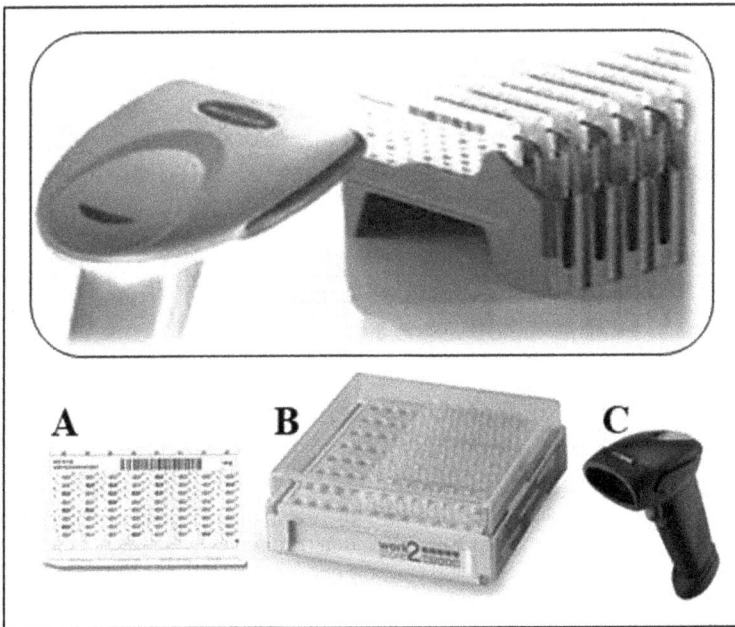

FIGURE 15.3   VITEK II compact cassette equipped with (A) 15 colorimetric reagent cards; (B) microbial suspension tubes; and (C) scanner device for enter results [11, 15].

15.2.1 *WORKING PRINCIPLE AND RESULTS*

The catalogs to identify VITEK system products are built using sets of well-described bacterial and yeast strains estimated under several culture conditions. These strains are isolated from several clinical, food, soil, and industrial samples as well as from microbial culture collection centers such as American type culture collection (ATCC) and cultural collections laboratory in universities [26–28]. After scanner process of colorimetric reagent cards, test response results show as + or –. Sometime the reactions results appear in parentheses as (+) or (-) that are indicator of weak responses. The results are matched with databases of the VITEK system as well as Bergey's Manual of Systematic Bacteriology as a certified reference [29–31]. After comparing the results of bio-pattern with VITEK system data, there is completion of numeral probability computation for all taxon. Several qualitative recognition scales are distributed based on the estimation of numerical probability computation [32, 33]. The various scales and related input are shown in Table 15.1. While studying the bio-pattern sample for an anonymous organism (e.g., bacteria, yeasts, etc.), the results are all fully negative or may depend on (i) negative testing and (ii) negative tests with false responses. The identification score will be "Non interactive bio-pattern." Viability and numbers of organisms is importance to get best identification of testing organisms [33, 34].

TABLE 15.1   The Levels of Microorganism's Identification [20]

Levels	Options	Probability percentage	Notes
Excellent	1 (One)	96.00–99.00%	–
Very good	1 (One)	93.00–95.00%	–
Good	1 (One)	89.00–92.00%	–
Accept	1 (One)	85.00–88.00%	–
Minimum identifies	2–3 (Two-three)	Summation/aggregate of choices = 100 after a single percent likelihood resolution represents the number of the option	2–3 species show the same bio-pattern separate by additional experimentations
Unidentified microbial	0 (Zero)	–	Either more three species show the same bio-pattern

15.3   APPLICATION IN IDENTIFICATION AND CLASSIFICATION

15.3.1   BACTERIA

Bacteria are classified and specified to differentiate between isolates to collect them from microbiologists and other scholars [4, 35]. Classification science seeks to characterize the variety of bacterial species using nomenclature and collecting them based on likes [36, 37]. Bacterial species can be classified on the base of cell structure, cellular metabolism, or on variances in bacterial cell components such as DNA, proteins, fatty acids, phospholipids pigments, antigens, and quinones. Several methods have been used to classify bacterial species such as biochemical tests and genetic methods [38, 39]. In past studies, *E. coli* O157:H7 bacterium was identified using selective media such as Sorbitol MacConkey agar and using biochemical tests. About 25% of isolates was matched with *E. coli* O157:H7 [40] while 36 isolates were identified to be *Aeromonas hydrophila* bacteria depending on morphological and microscopic methods. Around 10 biochemical tests and *Aero gene* detection were performed by PCR technique [41]. *Vibrio* spp. was detected in Iraqi White soft cheese product, eight species of *Vibrio* bacteria was obtained and identified by selective media, microscopic, and biochemical tests [42].

VITEK II system was used to identify several species of bacteria like lactic acid bacteria (LAB), probiotics, and pathogens. This technique is characterized by fast work completion and accurate results [43]. With Rapid ID 32A System, VITEK II system and MALDI-TOF, the entire 50 *Clostridium difficile* isolates were certainly listed. The findings were observed right in 0, 2, and 17 isolates of VITEK II system, Rapid ID 32A and MALDI-TOF for 18 non-*Cl. difficile* isolates, respectively [44]. The past study assessed the new VITEK II system identification cards which use colorimetric reactive to detect Gram-positive and Gram-negative bacteria (GP and GN cards, respectively) compared to fluorimetric cards (ID-GPC and ID-GNB) [45]. The research included all 580 bacterial isolates and strains of 116 taxa. 249 gram-positive bacteria examined with two cards (ID-GPC and GP), 218 and 235 bacterial isolate and strains were correctly identified, respectively. Around 331 Gram-negative bacteria examined with two cards *viz.* ID-GNB and GN, 295 and 321 bacterial strains were correctly identified, respectively [45].

15.3.1.1   LACTIC ACID BACTERIA (LAB)

Lactic acid bacteria (LAB) are gram-positive and non-sporic bacteria except some bacteria (*Sporolactobacillus, cocci*, coccobacillus, or bacilli) which have below 53 mol% G+C DNA base content. Usually, they are non-respiratory and do not contain a catalase enzyme [46, 47]. The glucose is fermented by them into lactic acid, CO_2, acetate, aroma, and ethanol. LAB grows anaerobically but unlike other anaerobes, is "aerotolerant anaerobic" and can grow in the presence of O_2. While they do not have catalase, they do have superoxide dismutase and other possible approaches of detoxifying radicals in peroxidase enzymes [47, 48].

The term "Lactic acid bacteria"-LAB is particularly used to denote the bacteria of order Lactobacillales which involves *Lactobacillus*, *Pediococcus*, *Leuconostoc*, *Lactococcus*, and *Streptococcus* besides *Carnobacterium*, *Aerococcus*, *Enterococcus*, *Sporolactobacillus*, *Oeno-coccus*, *Tetragenococcus*, *Vagococcus*, and *Weisella*. However, several genera of bacteria produce lactic acid as a major or minor end-product of carbohydrates fermentation [47, 49]. The VITEK II Compact System also identified the clinical isolates of gram-positive bacilli and *Enterococcus faecalis* and *E. faecium* [50]. The results accuracy of VITEK II Compact system was 66% to identify the *Enterococcus* spp. compared to *E. faecalis* and *E. faecium* as depicted in Table 15.2 [51]. These scores are like to those achieved in a previous study in which VITEK II showed 87% total accuracy for the identification of *Enterococcus* spp. [52].

TABLE 15.2   Identification of Bacterial Isolates by VITEK II Compact System [51]

Bacterial Isolates	Number of Isolates	Tested Percentage
(A) *Enterococcus* spp.		
E. avium	5	80%
E. casseliflavus	6	83.3%
E. durans	1	0%
E. faecalis	24	100%
E. faecium	23	100%
E. gallinarum	8	50%
E. raffinosus	6	0%
(B) *Streptococcus* spp.		
S. anginosus	4	25%
S. bovis	11	45.4%
S. mitis	3	66.66%
S. morbillorum	1	0%
S. mutans	1	0%
S. sanguis	6	50%

Source: Reprinted with permission from Ref. [51]. © 2002 Springer Nature.

The VITEK II Compact system with automated GP card was employed for the identification of LAB isolated from sucuk samples-dried and fermented sausage in Turkey. A full of 129 LAB isolates were specified among which *Lb. plantarum* dominated (45.7%) followed by 10.9% of *Lb. curvatus* and 9.3% of *Lb. fermentum*. *P. pentosaceus* and *P. acidilactici* isolates were specified from genus *Pediococcus* [53]. *Weissella* genus is one such genera of LAB possessing microbes with features like *Lactobacillus*. Several methods have been used to classify these bacteria, including VITEK II Compact system and 19 species have been identified. From those, *Weissella confusa*, *W. viridescens* and *W. cibaria* are the single isolated species from human's origin, while other species were isolated from raw milk, fermented cereals, and vegetables sources [54]. In the past research, the automatic VITEK II system with GP67 cards identified the *Enterococcus* genus which was also used to re-identify the *Lactococcous garvieae* (>99% likelihood) for the additional speciation [55]. The microorganisms of kefir samples were classified and then recognized using VITEK II Compact automated ID/AST card and API System. Further, all kefir samples contain *Kocuria* spp. when incubated at 4C, however, the incubation at 20°C, 25°C and 30°C did not show any traces. Some other microorganisms like *Leuconostoc* spp. and *Micrococcus* spp. were detected in whole kefir groups [56]. The identification of *E. faecalis* Y17 and *Pediococcus pentosaceus* G11 by using VITEK II compact system. They have antibacterial activity against *A. hydrophila*, *Vibrio alginolyticus*, *V. parahaemolyticus*, *Staphylococcus aureus*, and *Streptococcus* spp. [57].

15.3.1.2   PROBIOTIC BACTERIA

Probiotic bacteria are commonly known as "live microorganisms," which while ingested or locally utilized in appropriate numbers showed the health benefits for the host as human or animals [35]. The main probiotic bacteria applied in dairy products have been strains of *Lactobacillus acidophilus* and *Bifidobacterium* spp., which are orderly added to fermented dairy products like yogurt because of their increased functional advantages to gastrointestinal system health of human and animals [58, 59, 74].

The special cards were used to diagnose *Bifidobacterium* species and diagnostic card for anaerobic bacteria is known as ANC Card [60]. The record of ANC card includes 63 genera of anaerobes and Corynebacteria among which 20 genera are registered in ANC database of VITEK II system: *Actinomyces, Arcanobacterium, Bacteroides, Bifidobacterium, Clostridium, Collinsella, Corynebacterium, Eggerthella, Eubacterium, Finegoldia, Fusobacterium, Lactobacillus, Microbacterium, Parvimonas, Peptoniphilus, Peptostreptococcus, Prevotella, Propionibacterium, Staphylococcus*, and *Veillonella* [61]. *Bifidobacterium* isolates were obtained from several sources including Human milk and feces of children aged 7–43 days using the selective media MRS NLLP. Confirmatory diagnosis for 8 bacterial isolates using VITEK II compact system stressed the origin of four of them to bacteria *Bifidobacterium bifidum* [62].

15.3.1.3   PATHOGENIC BACTERIA

Pathogenic bacteria are those microorganisms which can cause diseases after entry into human and animals' body [4]. They can spread through food, water, soil, air, and during body contact. Generally, bacteria are harmless and helpful but some species are pathogenic [3]. In previous study, the new VITEK II compact system (bioMérieux, France) equipped with automatic GNB card identified the rod, and Gram-negative bacteria. All 845 pathogenic strains were examined, showing 70 specific taxa relations to the family Enterobacteriaceae or no enteric bacilli, whereas at species level, 716 bacterial strains, with an additional 32 strains were identified from several manual tests such as oxidase, hemolysis, motility, indole reactions, and pigmentation. Only 7 strains were mis-identified and 10 were not identified [63].

The compact system of VITEK II with automated GPC and GNB identification cards, which uses the fluorescent read process, required up to 3 h for identification of 52 Gram-positive strains and 98 Gram-negative strains. The system covering an extended database of 115 Gram-positive and 135 Gram-negative taxa with a new designed GP card for Gram-positive and GN card for Gram-negative formats as per colorimetric detection [11]. The VITEK II system GN card estimated 562 bacterial isolates of common as well as rarely found species having Gram-negative and rod-shaped bacteria in addition to 154 non-fermenting glucose isolates. API 20 E and API 20 NE kits were used to identify bacterial strains. VITEK II program GN card accurately recognized 96.8% of isolates with a minor discrimination of 6.4%. Misidentifications showed at 3.0% but no Identification showed at 0.2% [64].

Total 273 isolates were obtained from models of various clinical samples (urine, stool, blood, and sputum) and swabs from the vagina, wounds, ear, and crag. Among these isolates, 30 isolates of *Klebsiella* genus were found from the results of VITEK II compact system. *Klebsiella* have been past diagnosed by using API 20E kit as *K. pneumoniae* (90%) and *K. oxytoca* (10%) when the results of genetic methods reported 80% of isolates belonged to the specie of *K. pneumoniae* and 20% of *K. oxytoca* [65].

15.3.2 YEASTS

Being a member of Fungus kingdom, yeasts are eukaryotic microorganisms with monocular cells. The initial yeast was found hundreds of millions of years ago, and at minimum 1,500 species are recognized at percent time [66]. They are estimated to 1% of all described species of the fungus kingdom. Diagnosis and classification of yeasts is made by well-known diagnostic methods such as biochemical tests and genetic method [67, 68]. *Saccharomyces* contains several species of yeasts which are considered very significant in food production. It is famous as the baker's yeast or brewer's yeast. They are monocular and saprotrophic fungi. One example is *Saccharomyces cerevisiae*, which is applied in producing wine, bread, beer, baker products and for human and animal health [69, 70]. YST card is applied to identify 15 yeast species and yeast-like microbes using the automatic VITEK II compact system [13]. The list of supposed species is shown in Table 15.3.

TABLE 15.3 Identification of 185 Yeast Isolates Using VITEK II Compact System [13]

Yeast Isolates	Number of Isolates	Number of Identification	Number of Non-Identification
Candida albicans	48	48	0
C. dubliniensis	22	22	0
C. glabrata	23	23	23
C. guilliermondii	7	6	1
C. inconspicua	2	0	2
C. kefyr	1	1	0
C. krusei	10	10	0

TABLE 15.3   *(Continued)*

Yeast Isolates	Number of Isolates	Number of Identification	Number of Non-Identification
C. lipolytica	2	1	1
C. lusitaniae	7	6	1
C. parapsilosis	22	21	1
C. pelliculosa	2	2	0
C. rugosa	7	5	2
C. tropicalis	10	10	0
C. zeylanoides	2	1	1
Cryptococcus neoformans	5	4	1
Geotrichum candidum	2	2	0
Malassezia furfur	1	0	1
Rhodotorula glutinis	1	1	0
Saccharomyces cerevisiae	10	10	0
Trichosporon spp.	1	0	1
Trichosporon mucoides	2	2	0

The identification of yeast was performed using recent colorimetric cards (YST), mainly the fluorometric cards (ID-YST) of yeast and both these cards perfectly identified the 19 species of 172, 161 and 144 isolates of yeast [71]. The potential of VITEK II compact system was determined by identifying the antifungal sensitivity of 32 yeast clinical isolates considering the reference of Clinical and Laboratory Standards Institute and European Committee on Antimicrobial Susceptibility Testing reference standards. These strains were correctly identified (100% accuracy) using the VITEK II compact system [72]. Using VITEK II compact system for isolates, two commercial strains were identified, where probability percentage was 90–94% of *Saccharomyces cerevisiae*. Since there is no database in VITEK II compact system, it may include *Saccharomyces boulardii* because the biochemical tests did not supply enough guide to distinguish between *S. boulardii* isolate and *S. cerevisiae* isolate [73].

15.4   CONCLUDING REMARKS

The VITEK II compact is an automated identification and classification method for microorganisms like bacteria and fungi. Outcomes of the published studies suggest that VITEK II produces very rapid, accurate, and reproducible scores. VITEK II compact demonstrates state-of-the-art phenotypic identification technology with its colorimetric reagent card, hardware, and software. It is successfully used to identify and classify numerous types of bacteria and yeast strains with compared to accuracy of various biochemical tests and genomic tools. Very limited research had been performed using this new system and hence demands further well-structured and planned investigations in future. Furthermore, database build up is also extremely required for comparison with other well-established methods and technologies particularly from microbiological perspective. However, it can be concluded that at clinical level the use of such novel strategy will reduce time than other diagnostic methods currently used in laboratories to identify pathogenic or virulent strains.

KEYWORDS

- **analytical technique**
- **antimicrobial sensitivity**
- **bacteria**
- **clostridium difficile**
- **identification methods**
- **VITEK II**
- **yeasts**

REFERENCES

1. Verma, D. K., & Srivastav, P. P., (2017). *Microorganisms in Sustainable Agriculture, Food and the Environment* (Vol. 1). Apple Academic Press: USA.
2. Verma, D. K., Patel, A. R., Sandhu, K. S., Baldi, A., & García, S., (2020b). *Biotechnical Processing in the Food Industry: New Methods, Techniques, and Applications.* Apple Academic Press; USA.

3. Verma, D. K., Patel, A. R., Srivastav, P. P., Mohapatra, B., & Niamah, A. K., (2020c). *Microbiology for Food and Health Technological Developments and Advances.* Apple Academic Press: USA.

4. Verma, D. K., Mohapatra, B., Kumar, C., Bajwa, N., Kiran, K. S., Kimmy, G., Baldi, A., et al., (2020a). Molecular techniques for detection of foodborne pathogens: *Salmonella* and *Bacillus cereus*. In: Verma, D. K., Patel, A. R., Srivastav, P. P., Mohapatra, B., & Niamah, A. K., (eds.), *Microbiology for Food and Health Technological Developments and Advances* (pp. 231–296). Apple Academic Press: USA.

5. Donay, J. L., Mathieu, D., Fernandes, P., Pregermain, C., Bruel, P., Wargnier, A., Casin, I., et al., (2004). Evaluation of the automated phoenix system for potential routine use in the clinical microbiology laboratory. *J. Clinic. Microbiol., 42*(4), 1542–1546.

6. Burnham, C. A. D., Dunne, Jr. W. M., Greub, G., Novak, S. M., & Patel, R., (2013). Automation in the clinical microbiology laboratory. *Clinic. Chem., 59*(12), 1696–1702.

7. Bourbeau, P. P., & Ledeboer, N. A., (2013). Automation in clinical microbiology. *J. Clinic. Microbiol., 51*, 1658–1665.

8. Marschal, M., Bachmaier, J., Autenrieth, I., Oberhettinger, P., Willmann, M., & Peter, S., (2017). Evaluation of the accelerate pheno system for fast identification and antimicrobial susceptibility testing from positive blood cultures in bloodstream infections caused by gram-negative pathogens. *J. Clinic. Microbiol., 55*, 2116–2126.

9. Van, B. A., Bachmann, T. T., Lüdke, G., Lisby, J. G., Kahlmeter, G., Mohess, A., Becker, K., et al., (2019). Developmental roadmap for antimicrobial susceptibility testing systems. *Nat. Rev. Microbiol., 17*(1), 51–62.

10. Van, B. A., Burnham, C. A. D., Rossen, J. W., Mallard, F., Rochas, O., & Dunne, W. M., (2020). Innovative and rapid antimicrobial susceptibility testing systems. *Nat. Rev. Microbiol.,* 1–13.

11. Pincus, D. H., (2006). Microbial identification using the bioMérieux Vitek® 2 system. In: Miller, M. J., (ed.), *Encyclopedia of Rapid Microbiological Methods* (pp. 1–32).

12. Eigner, U., Schmid, A., Wild, U., Bertsch, D., & Fahr, A. M., (2005). Analysis of the comparative workflow and performance characteristics of the VITEK 2 and phoenix systems. *J. Clinic. Microbiol., 43*(8), 3829–3834.

13. Meurman, O., Koskensalo, A., & Rantakokko-Jalava, K., (2006). Evaluation of VITEK 2 for identification of yeasts in the clinical laboratory. *Clinic. Microbiol Infec., 12*, 591–593.

14. Machen, A., Drake, T., & Wang, Y. F. W., (2014). Same day identification and full panel antimicrobial susceptibility testing of bacteria from positive blood culture bottles made possible by a combined lysis-filtration method with MALDI-TOF VITEK mass spectrometry and the VITEK 2 system. *PloS One, 9*, e87870.

15. Al Mijalli, S. H. S., (2017). Bacterial contamination of indoor air in schools of Riyadh, Saudi Arabia. *Air Water Borne Dis., 6*, 1–8.

16. Chun, K., Syndergaard, C., Damas, C., Trubey, R., Mukindaraj, A., Qian, S., Jin, X., Breslow, S., & Niemz, A., (2015). Sepsis pathogen identification. *J. Lab. Autom., 20*, 539–561.

17. Lowe, P., Haswell, H., & Lewis, K., (2006). Use of various common isolation media to evaluate the new VITEK 2 colorimetric GN card for identification of *Burkholderia pseudomallei*. *J. Clinic. Microbiol., 44*, 854–856.

18. Hsieh, W. S., Sung, L. L., Tsai, K. C., & Ho, H. T., (2009). Evaluation of the VITEK 2 cards for identification and antimicrobial susceptibility testing of non-glucose-fermenting gram-negative bacilli. *APMIS, 117*, 241–247.

19. Yasmon, A., Rosana, Y., Usman, D., Prilandari, L. I., & Hartono, T. S., (2020). Identification and phylogenetic analysis of *Corynebacterium diphtheriae* isolates from Jakarta, Indonesia based on partial *rpoB* gene. *Biodiversitas, 21*, 3070–3075.

20. Chandler, L. J., LaSala, P. R., & Whittier, S., (2016). Rapid devices and instruments for the identification of aerobic bacteria. *Manual of Commercial Methods in Clinical Microbiology: International Edition, 21*–55.

21. Nakasone, I., Kinjo, T., Yamane, N., Kisanuki, K., & Shiohira, C. M., (2007). Laboratory-based evaluation of the colorimetric VITEK-2 Compact system for species identification and of the Advanced Expert System for detection of antimicrobial resistances: VITEK-2 Compact system identification and antimicrobial susceptibility testing. *Diagn. Microbiol. Infec. Dis., 58*, 191–198.

22. Bidossi, A., Bortolin, M., Toscano, M., De Vecchi, E., Romanò, C. L., Mattina, R., & Drago, L., (2017). *In vitro com*parison between α-tocopheryl acetate and α-tocopheryl phosphate against bacteria responsible of prosthetic and joint infections. *PLoS One, 12*, e0182323.

23. Abdel-Aleem, H., Dishisha, T., Saafan, A., Abou, K. A. A., & Gaber, Y., (2019). Bio cementation of soil by calcite/aragonite precipitation using *Pseudomonas azotoformans* and *Citrobacter freundii* derived enzymes. *RSC Advances, 9*, 17601–17611.

24. Bradford, R., Manan, R. A., Daley, A. J., Pearce, C., Ramalingam, A., D'mello, D., Mueller, Y., et al., (2006). Coagulase-negative staphylococci in very-low-birth-weight infants: Inability of genetic markers to distinguish invasive strains from blood culture contaminants. *Eur J Clin Microbiol Infect Dis., 25*(5), 283–290.

25. Van, D. P. B., Klak, A., Desmet, S., & Verhaegen, J., (2018). How small modifications in laboratory workflow of blood cultures can have a significant impact on time to results. *Eur J Clin Microbiol Infect Dis., 37*, 1753–1760.

26. Jackson, E. E., & Forsythe, S. J., (2016). Comparative study of *Cronobacter* identification according to phenotyping methods. *BMC Microbiol., 16*(1), 146.

27. Pinheiro, D., Monteiro, C., Faria, M. A., & Pinto, E., (2019). Vitek® MS v3. 0 system in the identification of filamentous fungi. *Mycopathologia, 184*, 645–651.

28. Huber, C. A., Pflüger, V., Reed, S., Cottrell, K., Sidjabat, H. E., Ranasinghe, A., Zowawi, H. M., et al., (2019). Bacterial identification using a SCIEX 5800 TOF/TOF MALDI research instrument and an external database. *J. Microbiol. Methods, 164*, 105685.

29. Stuckey, S., (2007). Automated systems: An overview. In: Schwalbe, R., Steele-Moore, L., & Goodwin, A. C., (eds.), *Antimicrobial Susceptibility Testing Protocols* (pp. 5–10). CRC Press: Boca Raton.

30. Vos, P., Garrity, G., Jones, D., Krieg, N. R., Ludwig, W., Rainey, F. A., Whitman, W. B., (2011). The Firmicutes. In: *Bergey's Manual of Systematic Bacteriology* (Vol. 3). Springer Science & Business Media.

31. Goodfellow, M., Kämpfer, P., Busse, H. J., Trujillo, M. E., Suzuki, K. I., Ludwig, W., & Whitman, W. B., (2012). *Bergey's Manual® of Systematic Bacteriology: Volume Five the Actinobacteria, Part A* (pp. 171–206). Springer: New York.

32. Singh, V., Padmanabhan, P., & Saha, S., (2019). Evaluations of the VITEK 2 BCL card for identification of biosurfactant producing bacterial isolate SPS1001. In: *Biotechnology and Biological Sciences: Proceedings of the 3rd International Conference of Biotechnology and Biological Sciences* (p. 315). CRC Press: Kolkata, India.

33. Adekoya, I., Obadina, A., Olorunfemi, M., Akande, O., Landschoot, S., De Saeger, S., & Njobeh, P., (2019). Occurrence of bacteria and endotoxins in fermented foods and beverages from Nigeria and South Africa. *Inter. J. Food Microbiol., 305*, 108251.

34. Darbandi, F., (2011). *Parallel Comparison of Accuracy in VITEK 2 Auto Analyzer and API 20 E/API 20 NE Microsystems.* MSc dissertation, University College of Borås, Borås.

35. Verma, D. K., Patel, A., Prajapati, J. B., Thakur, M., Al-Manhel, A. J. A., & Srivastav, P. P., (2020d). Starter culture and probiotic bacteria in dairy food products. In: Verma, D. K., Patel, A. R., Srivastav, P. P., Mohapatra, B., & Niamah, A. K., (eds.), *Microbiology for Food and Health Technological Developments and Advances* (pp. 2–42). Apple Academic Press: USA.

36. Woese, C. R., Blanz, P., & Hahn, C. M., (1984). What isn't a pseudomonad: The importance of nomenclature in bacterial classification. *Syst. Appl. Microbiol., 5*, 179–195.

37. Verma, D. K., Patel, A. R., & Srivastav, P. P., (2018). *Bioprocess Technology in Food and Health: Potential Applications and Emerging Scope.* Apple Academic Press: USA.

38. Jordan, J. A., Butchko, A. R., & Durso, M. B., (2005). Use of pyrosequencing of 16S rRNA fragments to differentiate between bacteria responsible for neonatal sepsis. *J. Mol. Diagn., 7*, 105–110.

39. Patel, A., Prajapati, J. B., & Nair, B. M., (2012). Methods for isolation, characterization and Identification of probiotic bacteria to be used in functional foods. *Int. J. Fermented Foods, 1*, 1–13.

40. Al-Kuzayi, A. K. N., & Al-Sahlany, S. T. G., (2011). Detecting for *E. coli* O157: H7 in dairy products which where locally processed and found in Basra city markets. *Basrah J. Agric. Sci., 24*, 290.

41. Niamah, A. K., (2012). Detected of aero gene in *Aeromonas hydrophila* isolates from shrimp and peeled shrimp samples in local markets. *J. Microbiol. Biotechnol. Food Sci., 2*, 634–639.

42. Al-Sahlany, S. T. G., (2016). Effect of *Mentha piperita* essential oil against *Vibrio* spp. isolated from local cheeses. *Pakistan J. Food Sci., 26*, 65–71.

43. Silva, L. F., Casella, T., Gomes, E. S., Nogueira, M. C. L., De Dea, L. J., & Penna, A. L. B., (2015). Diversity of lactic acid bacteria isolated from Brazilian water buffalo mozzarella cheese. *J. Food Sci., 80*, M411–M417.

44. Kim, Y. J., Kim, S. H., Park, H. J., Park, H. G., Park, D., Am Song, S., Lee, H. J., et al., (2016). MALDI-TOF MS is more accurate than VITEK II ANC card and API Rapid ID 32 A system for the identification of *Clostridium* species. *Anaerobe, 40*, 73–75.

45. Wallet, F., Loïez, C., Renaux, E., Lemaitre, N., & Courcol, R. J., (2005). Performances of VITEK 2 colorimetric cards for identification of gram-positive and gram-negative bacteria. *J. Clinic. Microbiol., 43*, 4402–4406.

46. Fritze, D., & Claus, D., (1995). Spore-forming, lactic acid producing bacteria of the genera *Bacillus* and *Sporolactobacillus*. In: *The Genera of Lactic Acid Bacteria* (pp. 368–391). Springer: Boston, MA.

47. Patel, A., Shah, N., & Verma, D. K., (2017). Lactic acid bacteria (LAB) bacteriocins: An ecological and sustainable biopreservative approach to improve the safety and shelf-life of foods. In: Verma, D. K., & Srivastav, P. P., (eds.), *Microorganisms in Sustainable Agriculture, Food and the Environment* (pp. 197–258). Apple Academic Press; USA.

48. Condon, S., (1983). Aerobic metabolism of lactic acid bacteria. *Irish J. Food Sci. Technol., 7*(1), 15–25.

49. Makarova, K., Slesarev, A., Wolf, Y., Sorokin, A., Mirkin, B., Koonin, E., Pavlov, A., et al., (2006). Comparative genomics of the lactic acid bacteria. *Proc. Natl. Acad. Sci., 103*, 15611–15616.

50. Brigante, G., Luzzaro, F., Bettaccini, A., Lombardi, G., Meacci, F., Pini, B., Stefani, S., & Toniolo, A., (2006). Use of the phoenix automated system for identification of *Streptococcus* and *Enterococcus* spp. *J. Clinic. Microbiol., 44*(9), 3263–3267.

51. Gavin, P. J., Warren, J. R., Obias, A. A., Collins, S. M., & Peterson, L. R., (2002). Evaluation of the VITEK 2 system for rapid identification of clinical isolates of gram-negative bacilli and members of the family *Streptococcaceae*. *Eur. J. Clin. Microbiol. Infect. Dis., 21*(12), 869–874.

52. Garcia-Garrote, F., Cercenado, E., & Bouza, E., (2000). Evaluation of a new system, VITEK 2, for identification and antimicrobial susceptibility testing of enterococci. *J. Clinic. Microbiol., 38*(6), 2108–2111.

53. Kaban, G., & Kaya, M., (2008). Identification of lactic acid bacteria and gram-positive catalase-positive cocci isolated from naturally fermented sausage (sucuk). *J. Food Sci., 73*, M385–M388.

54. Kamboj, K., Vasquez, A., & Balada-Llasat, J. M., (2015). Identification and significance of *Weissella* species infections. *Front Microbiol., 6*, 1204.

55. Choksi, T. T., & Dadani, F., (2017). Reviewing the emergence of *Lactococcus garvieae*: A case of catheter associated urinary tract infection caused by *Lactococcus garvieae* and *Escherichia coli* coinfection. *Cas. Rep. Infec. Dis.,* 5921865.

56. Hecer, C., Ulusoy, B., & Kaynarca, D., (2019). Effect of different fermentation conditions on composition of kefir microbiota. *Inter. Food Res. J., 26*(2), 401–409.

57. Yang, Q., Lü, Y., Zhang, M., Gong, Y., Li, Z., Tran, N. T., He, Y., et al., (2019). Lactic acid bacteria, *Enterococcus faecalis* Y17 and *Pediococcus pentosaceus* G11, improved growth performance, and immunity of mud crab (*Scylla paramamosain*). *Fish Shellfish Immunol., 93*, 135–143.

58. Niamah, A. K., Al-Sahlany, S. T. G., & Al-Manhel, A. J., (2016). Gum Arabic uses as prebiotic in yogurt production and study effects on physical, chemical properties and survivability of probiotic bacteria during cold storage. *World Appl. Sci. J., 34*(9), 1190–1196.

59. Niamah, A. K., & Verma, D. K., (2017). Microbial intoxication in dairy food product. In: Verma, D. K., & Srivastav, P. P., (eds.), *Microorganisms in Sustainable Agriculture, Food and the Environment* (pp. 143–170). Apple Academic Press: USA.

60. Rosenblatt, J. E., (1985). Anaerobic bacteria. In: *Laboratory Procedures in Clinical Microbiology* (pp. 315–378). Springer: New York.
61. Lee, E. H. L., Degener, J. E., Welling, G. W., & Veloo, A. C. M., (2011). Evaluation of the VITEK 2 ANC card for identification of clinical isolates of anaerobic bacteria. *J. Clinic. Microbiol., 49*, 1745–1749.
62. Al-Zubidi, M. A., (2016). *Isolation and Identification of Bifidobacterium Bifidum and Study the Effect of Metabolic Products in Bio Preservation of Some Meat Products.* MSc Dissertation, University of Basrah, Iraq.
63. Funke, G., Monnet, D., De Bernardis, C., Von, G. A., & Freney, J., (1998). Evaluation of the VITEK 2 system for rapid identification of medically relevant gram-negative rods. *J Clinic. Microbiol., 36*(7), 1948–1952.
64. Chatzigeorgiou, K. S., Sergentanis, T. N., Tsiodras, S., Hamodrakas, S. J., & Bagos, P. G., (2011). Phoenix 100 versus VITEK 2 in the identification of gram-positive and gram-negative bacteria: A comprehensive meta-analysis. *J. Clinic. Microbiol., 49*(9), 3284–3291.
65. AL-Talabany, B. M., AL-Abdeen, S. S. Z., & AL-Jobory, I. S., (2017). Comparative diagnostic study of *Klebsiella* using Api20E System and the device VITEK 2 and PCR. *Kirkuk University Journal/Scientific Studies, 12,* 81–95.
66. Hawksworth, D. L., & Lücking, R., (2017). Fungal diversity revisited: 2.2 to 3.8 million species. In: Heitman, J., Howlett, B. J., Crous, P. W., Stukenborck, E. H., James, T. Y., & Gow, N. A. R., (eds.), *The Fungal Kingdom* (pp. 79–95). ASM Pres.
67. Fleet, G. H., (1990). Yeasts in dairy products. *J. Appl. Bacteriol., 68*(3), 199–211.
68. Kurtzman, C., Fell, J. W., & Boekhout, T., (2011). *The Yeasts: A Taxonomic Study.* Elsevier.
69. Niamah, A. K., (2017). Physicochemical and microbial characteristics of yogurt with added *Saccharomyces boulardii. Curr. Nutr. Food Sci. J., 5*, 300–307.
70. Al-Sahlany, S. T. G., Altemimi, A. B., Al-Manhel, A. J. A., Niamah, A. K., Lakhssassi, N., & Ibrahim, S. A., (2020). Purification of bioactive peptide with antimicrobial properties produced by *Saccharomyces cerevisiae. Foods, 9*, 324.
71. Loïez, C., Wallet, F., Sendid, B., & Courcol, R. J., (2006). Evaluation of VITEK 2 colorimetric cards versus fluorimetric cards for identification of yeasts. *Diagn Microbiol. Infect. Dis., 56*, 455–457.
72. Melhem, M. S. C., Bertoletti, A., Lucca, H. R. L., Silva, R. B. O., Meneghin, F. A., & Szeszs, M. W., (2013). Use of the VITEK 2 system to identify and test the antifungal susceptibility of clinically relevant yeast species. *Braz. J. Microbiol., 44*, 1257–1266.
73. AL-Zaiadi, R., AL-Shekdhaher, A., & AL-Jelaw, M., (2016). Isolation and identification of food probiotic *Saccharomyces boulardii* by using traditional methods, VITEK 2 system and molecular identification methods. *Iraq Journal of Market Research and Consumer Protection, 8*, 42–60.
74. Niamah, A. K., Sahi, A. A., & Al-Sharifi, A. S., (2017). Effect of feeding soy milk fermented by probiotic bacteria on some blood criteria and weight of experimental animals. *Probiotics Antimicrob. Proteins, 9*, 284–291.

Index

For Product Safety Concerns and Information please contact our EU
representative GPSR@taylorandfrancis.com
Taylor & Francis Verlag GmbH, Kaufingerstraße 24, 80331 München, Germany